VOLUME SIXTY

ADVANCES IN INSECT PHYSIOLOGY

Genes and Endocrine Signalling in Development and Homeostasis

VOLUME SIXTY

ADVANCES IN INSECT PHYSIOLOGY

Genes and Endocrine Signalling in Development and Homeostasis

Edited by

MICHAEL E. ADAMS

Department of Entomology;
Department of Molecular, Cell, and Systems Biology, University of California, Riverside, CA, United States

ELSEVIER

ACADEMIC PRESS

An imprint of Elsevier

Academic Press is an imprint of Elsevier
125 London Wall, London, EC2Y 5AS, United Kingdom
The Boulevard, Langford Lane, Kidlington, Oxford OX5 1GB, United Kingdom
525 B Street, Suite 1650, San Diego, CA 92101, United States
50 Hampshire Street, 5th Floor, Cambridge, MA 02139, United States

First edition 2021

ISBN: 978-0-323-85425-2
ISSN: 0065-2806

For information on all Academic Press publications
visit our website at https://www.elsevier.com/books-and-journals

Publisher: Zoe Kruze
Acquisitions Editor: Sam Mahfoudh
Developmental Editor: Naiza Ermin Mendoza
Production Project Manager: Abdulla Sait
Cover Designer: Victoria Pearson

Typeset by SPi Global, India

Contents

Contributors

Michael E. Adams
Department of Entomology; Department of Molecular, Cell, and Systems Biology, University of California, Riverside, CA, United States

Lenka Bittova
Biology Center of the Czech Academy of Sciences, Institute of Entomology, Ceske Budejovice, Czech Republic

Juan Du
Department of Entomology and MOA Key Lab of Pest Monitoring and Green Management, College of Plant Protection, China Agricultural University, Beijing, China

Tiantian Fu
Department of Entomology and MOA Key Lab of Pest Monitoring and Green Management, College of Plant Protection, China Agricultural University, Beijing, China

Lei He
Department of Entomology and MOA Key Lab of Pest Monitoring and Green Management, College of Plant Protection, China Agricultural University, Beijing, China

Qiankun He
Department of Entomology and MOA Key Lab of Pest Monitoring and Green Management, College of Plant Protection, China Agricultural University, Beijing, China

Marek Jindra
Biology Center of the Czech Academy of Sciences, Institute of Entomology, Ceske Budejovice, Czech Republic

Yahong Li
Department of Entomology and MOA Key Lab of Pest Monitoring and Green Management, College of Plant Protection, China Agricultural University, Beijing, China

Zhe Liu
Department of Entomology and MOA Key Lab of Pest Monitoring and Green Management, College of Plant Protection, China Agricultural University, Beijing, China

Pengfei Lv
Department of Entomology and MOA Key Lab of Pest Monitoring and Green Management, College of Plant Protection, China Agricultural University, Beijing, China

Matej Milacek
Department of Molecular Biology and Genetics, University of South Bohemia, Ceske Budejovice, Czech Republic

Sarka Tumova
Biology Center of the Czech Academy of Sciences, Institute of Entomology, Ceske Budejovice, Czech Republic

Yu Wei
Department of Entomology and MOA Key Lab of Pest Monitoring and Green Management, College of Plant Protection, China Agricultural University, Beijing, China

Naoki Yamanaka
Department of Entomology, University of California, Riverside, CA, United States

Zhangwu Zhao
Department of Entomology and MOA Key Lab of Pest Monitoring and Green Management, College of Plant Protection, China Agricultural University, Beijing, China

Preface

Volume 60 of the *Advances in Insect Physiology* series devotes four chapters to increased understanding of how genes and endocrine-signalling molecules contribute to physiological changes and stability in insects. Homeostasis as a concept describes maintenance of a steady state within or between cells, organs, and organisms. Its Greek roots are "homoios-" meaning "similar" and "stasis" referring to "steady". Homeostatic regulation can extend to programming of body form, daily activity and sleep rhythms, temperature regulation, ion and water balance, reproductive state, synaptic transmission, and many other processes. Among social insects, the term also can refer to maintenance of cooperative interactions in pursuit of stability within the colony.

Hormones have important roles in achievement of steady state after rapid growth periods. For example, during ontogeny, moulting that permits growth is promoted by ecdysteroids, while juvenile hormones (JHs) maintain the "status quo" larval body form. Upon reaching critical body weight and nutritional status, steroid signalling in the absence of JH leads to pupal commitment and onset of metamorphosis. Post-metamorphosis, JH and 20E are re-purposed as homeostatic regulators of reproduction, associated sex-linked behavioural traits, and daily activity rhythms of locomotory activity and sleep.

In Chapters 1 and 2, Yamanaka and Jindra et al. summarize the nature and actions of the two principle lipophilic hormones in insects: 20E and JH. Whereas the signalling molecules themselves were discovered and identified in the 1960s, ecdysone receptors were characterized only in the last 30 years and JH receptors within the last 10 years. Both JH and 20E receptors are transcription factors that regulate gene expression. Identification of the JH receptor overcame years of uncertainty surrounding the role of the *Methoprene-tolerant* (Met) gene of *Drosophila*, whose deficiency failed to cause characteristic morphogenetic defects during development expected from classical experiments in *Rhodnius* and lepidopterans. This issue was finally resolved through use of RNA interference in complementary insect models, principally with the flour beetle *Tribolium*. These works confirmed the identity of Met (and also *Drosophila* gce) as the authentic JH receptor responsible for both the developmental and reproductive actions of JH as well as the growth regulator (disruptor) actions of juvenoids like methoprene, fenoxycarb, and pyriproxyfen. Armed with Met, defined as the first ever bHLH-PAS transcription factor functioning as a hormone receptor, Jindra and others are spearheading ongoing studies aimed at a deeper understanding of ligand-receptor interactions, nucleocytoplasmic transport, and how posttranslational modifications of the receptor influence JH signalling. Met is also now employed as a target for large-scale screening for new insect control agents with greater target organism selectivity and reduced environmental impacts.

Yamanaka's review provides a comprehensive overview of the complex pleiotropic actions of ecdysteroids, including moulting, metamorphosis, reproduction, and stress responses. In covering aspects of biosynthesis, secretion, transport, structural modification, and circulation, some recent and unexpected research highlights stand out. One of these involves secretion. While classification of 20E as a lipophilic steroid hormone has led to the standard textbook presumption that these molecules pass freely through cell membranes, it is now clear that 20E is packaged into secretory vesicles and released via conventional exocytosis just as neurotransmitters and peptide hormones. If ecdysteroids cannot freely leave cells, how do they enter on the way to intracellular receptor binding? More recent work points to "facilitated diffusion" through membranes via transporters. These advances will have consequences for how steroid hormones move in and out of cells and how future insect control strategies may be developed.

Chapter 3 summarizes the past decade of research on the epitracheal endocrine system, drawing from studies dating back to its first characterization by Zitnan et al. in 1996 for context. Endocrine Inka cells synthesize and store ecdysis triggering hormones (ETHs) during the early wave of gene expression that initiates moulting. Declining steroid levels and late gene expression confer Inka cell secretory competence to release ETHs. Upon ETH release, a cascade of endocrine signals regulates tracheal air-filling and sequential ecdysis motor patterns to promote successful transition to the next stage. Recent reports demonstrate obligatory involvement of multiple peptide hormones, including eclosion hormone, CCAP, bursicon, kinin, myoinhibitory peptides, and others. New insights into feedback and feed-forward signalling have been recently described, lending a deeper understanding of how ETH triggers multiple layers of neuronal activity to facilitate ecdysis. Following metamorphosis, ETH is re-purposed for adult functions, including promotion of vitellogenesis, ovulation, and adaptation to stress. Finally, recent evidence demonstrates that ETH along with JH is crucial for maintenance of short- and long-term courtship memories in flies.

Many of these regulatory functions during adulthood appear to involve a signalling triad involving 20E, ETH, and JH.

In Chapter 4, Zhao et al. provide a broad perspective on sleep research in *Drosophila*, summarizing how circadian rhythms, diverse signalling molecules, intracellular signal transduction pathways, and gene regulation influence sleep/wake cycles. This chapter highlights the usefulness of the genetically malleable fruit fly in unravelling the homeostatic regulation of sleep and how aberrant regulation may factor into a number of disease states.

Michael E. Adams
Department of Entomology;
Department of Molecular, Cell, and Systems Biology,
University of California, Riverside, CA, United States

CHAPTER ONE

Ecdysteroid signalling in insects—From biosynthesis to gene expression regulation

Naoki Yamanaka*
Department of Entomology, University of California, Riverside, CA, United States
*Corresponding author: e-mail address: naoki.yamanaka@ucr.edu

Contents

Abstract

Ecdysteroids are a group of steroid hormones originally identified as moulting hormones in insects. Ever since the isolation of the first ecdysteroid, studies in multiple model insect species have revealed their modes of action and pleiotropic functions. It is now established that they control a multitude of physiological processes, including moulting, metamorphosis, reproduction, and stress responses. Here I summarize our current understanding of the diverse functions of ecdysteroids throughout the insect life cycle. I then highlight six steps of ecdysteroid signalling as a framework to comprehend this complex hormone signalling pathway: biosynthesis, secretion, circulation, transport, modification, and gene expression regulation.

Advances in Insect Physiology, Volume 60
ISSN 0065-2806
https://doi.org/10.1016/bs.aiip.2021.03.002

1. Introduction

Early studies by insect physiologists revealed that insect moulting is under endocrine control (Fukuda, 1940; Kopeć, 1922; Wigglesworth, 1934; Williams, 1947). In 1954, the first "moulting hormone" was isolated and named ecdysone (Butenandt and Karlson, 1954), followed by identification of the second hormone, 20-hydroxyecdysone (20E), in 1966 (Hampshire and Horn, 1966). Ecdysone and 20E are steroid hormones with multiple hydroxyl groups on the tetracyclic steroidal backbone, as well as on the side chain (Fig. 1). Today, this unique family of steroids is collectively called ecdysteroids (Goodwin et al., 1978; Lafont et al., 2012), and hundreds of ecdysteroids with related structures have now been reported (http://ecdybase.org). Interestingly, many of them are found in plants that likely use these moulting hormone analogues (termed phytoecdysteroids) to deter phytophagous insects (Dinan, 2009; Dinan et al., 2009; Soriano et al., 2004).

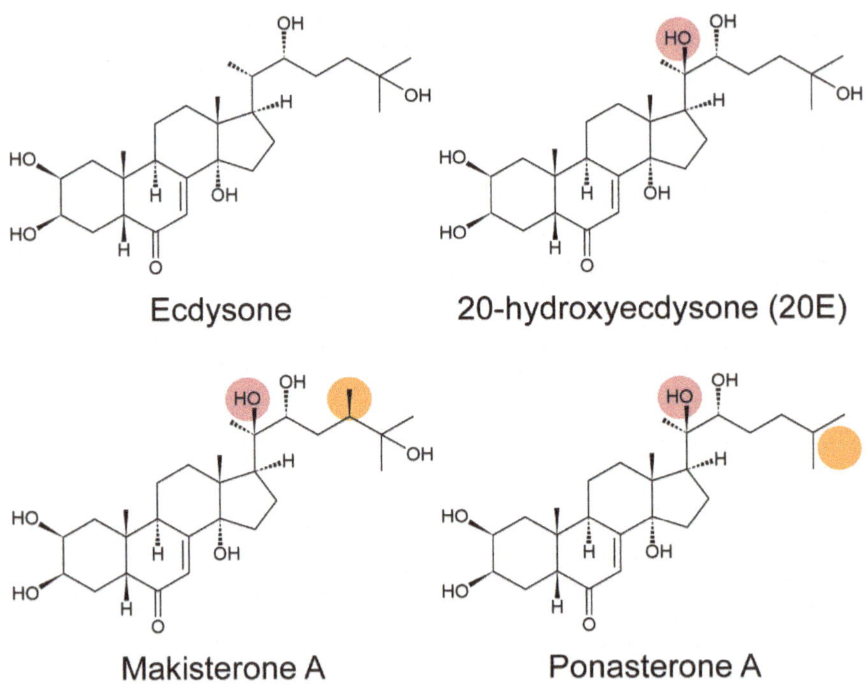

Fig. 1 Chemical structures of representative ecdysteroids. Structural differences from ecdysone are marked by circles. 20 carbon hydroxylation (pink circles) is the critical activation step for ecdysteroids.

Ever since the discovery of the first ecdysteroid, the simple body structure and life history of insects have greatly contributed to the advancement of steroid hormone research, as exemplified by the first report of the genomic action of steroid hormones in flies (Clever and Karlson, 1960) and multiple key discoveries related to nuclear receptor functions (Evans and Mangelsdorf, 2014; King-Jones and Thummel, 2005). Recent application of molecular genetic tools, particularly those of the fruit fly *Drosophila melanogaster*, has further led to numerous important findings in the past few decades (Pan et al., 2020; Yamanaka et al., 2013). In this chapter, I first review pleiotropic functions of ecdysteroids during the life cycle of insects. I then highlight six key steps in ecdysteroid signalling, with particular focus on recent findings in *D. melanogaster*, aiming to understand the diverse functions of ecdysteroids in a comprehensive manner.

2. Pleiotropic functions of ecdysteroids during the insect life cycle

2.1 Cuticle formation and morphogenesis during embryogenesis

Most of the so-called "Halloween" genes that encode ecdysteroidogenic enzymes were identified from the screen for mutations affecting the cuticle structure in embryos of *D. melanogaster* (Jürgens et al., 1984; Nüsslein-Volhard et al., 1984; Wieschaus and Nüsslein-Volhard, 2016; Wieschaus et al., 1984). They are called as such because of the somewhat bizarre appearance of mutant embryos exhibiting undifferentiated cuticle (Chávez et al., 2000; Gilbert, 2004). The Halloween mutants also display defective morphogenesis during embryogenesis, including defects in head involution, dorsal closure, gut formation, and nervous system development (Chávez et al., 2000; Giesen et al., 2003). These mutant phenotypes indicate critical functions of ecdysteroids in controlling cuticle deposition and morphogenesis during embryonic development.

2.2 Moulting during post-embryonic development

Insect moulting involves a series of events beginning with apolysis (cuticle detachment from the underlying epidermis) and ending with ecdysis (shedding of the old cuticle). Due to the systemic nature of these events, ecdysteroids and some other hormones coordinate cellular processes that accompany each step. At the beginning of moulting, apolysis is induced by rising levels of ecdysteroids in the haemolymph (Agui, 1977; Riddiford, 1985). Along with the rise of the ecdysteroid titre, the new cuticle layer is deposited by the epidermis, which continues as the ecdysteroid titre declines (Charles, 2010). The decline of the ecdysteroid titre induces expression of the nuclear receptor βFTZ-F1 in epitracheal endocrine Inka cells, which leads to the release of ecdysis-triggering hormone (ETH) (Cho et al., 2014; Kingan and Adams, 2000; Truman, 2005; Žitňan and Adams, 2012; Žitňan et al., 1996, 1999). ETH along with eclosion hormone act on multiple ensembles of peptidergic neurons in the central nervous system, which together coordinate ecdysis behaviour (Diao et al., 2016, 2017; Kim et al., 2006a,b; Mena et al., 2016; Scott et al., 2020). Finally, at the end of moulting, sclerotization and melanization of the new cuticle are initiated by downstream components of the ecdysteroid signalling cascade (Flaven-Pouchon et al., 2020; Hiruma and Riddiford, 2009).

2.3 Tissue degradation, remodelling, and morphogenesis during metamorphosis

In holometabolous insects, most larval tissues undergo histolysis during metamorphosis in response to ecdysteroid signalling (Nicolson et al., 2015; Xu et al., 2020b). Ecdysteroids induce programmed cell death either through caspase-mediated apoptosis or autophagy in larval tissues such as the salivary gland, midgut, and muscle (Aizenzon and Zhimulev, 1975; Denton et al., 2009; Jiang et al., 1997; Lee and Baehrecke, 2001; Lee et al., 2002a,b; Schwartz and Truman, 1982; Weeks and Truman, 1985; Zirin et al., 2013). In these tissues, expression of pro-apoptotic/autophagy genes are induced either directly or indirectly by ecdysteroids (Cakouros et al., 2004; Denton et al., 2013; Gorski et al., 2003; Jiang et al., 2000; Kang and Bashirullah, 2014; Lee et al., 2003). Along with removal of obsolete larval tissues, ecdysteroids also induce cell differentiation and morphogenesis in adult-specific tissues during metamorphosis, including various external structures that develop from imaginal discs, such as compound eyes, antennae, legs, and wings. Indeed, mitogenic effects of ecdysteroids on these tissues can be observed from larval stages, whereas high ecdysteroid titres during metamorphosis further induce cell differentiation and tissue morphogenesis (Champlin and Truman, 1998; Dye et al., 2017; Gancz et al., 2011; Herboso et al., 2015; Hodin and Riddiford, 1998; Kato and Riddiford, 1987; Kawasaki, 1995; Koyama et al., 2004; Mendes and Mirth, 2016; Mirth et al., 2009; Nijhout and Grunert, 2002; Nijhout et al., 2007, 2018). Ecdysteroids interact with several growth regulatory pathways in developing adult tissues to induce morphogenesis (Ninov et al., 2009;

Verma and Cohen, 2015; Zhang et al., 2015). Moreover, they trigger systemic metabolic changes at the beginning of metamorphosis, including upregulation of insulin-like peptide production in the fat body (Okamoto et al., 2009a,b; Slaidina et al., 2009), as well as the induction of a metabolic shift in energy homeostasis (Nishimura, 2020; Yamada et al., 2020), to further support tissue growth during metamorphosis.

In the nervous system, neuronal networks are reorganized during metamorphosis to serve adult-specific functions such as mating and flight (Tissot and Stocker, 2000; Truman, 1996; Yaniv and Schuldiner, 2016). This remodelling happens through a series of ecdysteroid-dependent processes, starting with cell fate specification of neural progenitor cells during the larval stage (Dillard et al., 2018; Lanet et al., 2013; Marchetti and Tavosanis, 2019; Syed et al., 2017a,b), followed by programmed cell death of larval neurons (Choi et al., 2006; Hara et al., 2013; Robinow et al., 1993, 1997; Truman and Schwartz, 1984; Wang et al., 2019; Winbush and Weeks, 2011), remodelling of axonal as well as dendritic arbours (Awasaki and Lee, 2011; Kuo et al., 2006; Lee et al., 2000; Schubiger et al., 1998; Truman and Reiss, 1976; Williams and Truman, 2005; Yu and Schuldiner, 2014), and differentiation of adult-specific neurons (Brown and Truman, 2009; Veverytsa and Allan, 2012; Zhang et al., 2018a) during the pupal stage. Ecdysteroids thus coordinate multiple cellular events during metamorphosis in preparation for the adult stage.

2.4 Reproduction and stress responses during the adult stage

Ecdysteroids are known to regulate female reproduction in a variety of insects, including dipterans (Audit-Lamour and Busson, 1981; Hagedorn and Fallon, 1973), lepidopterans (Chatani and Ohnishi, 1976; Swevers and Iatrou, 2003), hymenopterans (Bloch et al., 2000; Geva et al., 2005), coleopterans (Parthasarathy et al., 2010), and dictyopterans (Hult et al., 2015). Detailed functions of ecdysteroid signalling in female reproduction are particularly well studied using dipteran model species, such as *D. melanogaster* and the yellow fever mosquito *Aedes aegypti* (Belles and Piulachs, 2015; Roy et al., 2018; Swevers, 2019; Uryu et al., 2015). During oogenesis, ecdysteroids regulate a variety of steps, including germline stem cell proliferation and differentiation (Ables and Drummond-Barbosa, 2010; Ables et al., 2016; Ameku and Niwa, 2016; Gancz et al., 2011; König et al., 2011; Yoshinari et al., 2019), formation of the follicular epithelium (Romani et al., 2009), initiation and progression of vitellogenesis

(Buszczak et al., 1999; Hagedorn, 2005; Raikhel et al., 2005; Richard et al., 1998; Sieber and Spradling, 2015; Terashima and Bownes, 2004; Terashima et al., 2005), eggshell formation including choriogenesis and micropyle formation (Bai et al., 2000; Bernardi et al., 2009; Hackney et al., 2007; Manning et al., 2017), and ovulation (Knapp and Sun, 2017). Ecdysteroids also regulate synthesis of sex pheromones (Adams et al., 1984; Blomquist et al., 1984, 1995; Tillman et al., 1999; Wicker and Jallon, 1995), sexual behaviours (Ganter et al., 2012), as well as the post-mating increase of food intake (Ahmed et al., 2020; Hadjieconomou et al., 2020; Sieber and Spradling, 2015; Zipper et al., 2020) in adult females. Although they also act on male reproductive organs (Fujinaga et al., 2017; Sharma et al., 2017; Shinbo and Happ, 1989; Xu et al., 2012) and regulate male sexual behaviours (Bozzolan et al., 2015; Dalton et al., 2009; Duportets et al., 2013; Ganter et al., 2007, 2011), ecdysteroid functions in male reproduction are much less well understood as compared to their roles in regulation of female reproduction.

Recent studies indicate that ecdysteroid signalling is upregulated in response to various environmental stressors such as high temperature, nutrient shortage, courtship deprivation, and sleep deprivation in adult insects (Hirashima et al., 2000; Ishimoto and Kitamoto, 2010; Ishimoto et al., 2009; Meiselman et al., 2018; Terashima et al., 2005). Elevated ecdysteroid signalling in turn modulates reproductive physiology (Gruntenko et al., 2005; Ishimoto et al., 2009; Meiselman et al., 2018; Terashima et al., 2005), as well as sleep homeostasis and innate immunity (Flatt et al., 2008; Han et al., 2017; Ishimoto and Kitamoto, 2010; Regan et al., 2013; Reynolds et al., 2020; Rus et al., 2013; Sun et al., 2016; Zhang and Palli, 2009). Ecdysteroids thus serve as stress hormones affecting multiple aspects of adult physiology, reminiscent of glucocorticoid functions in mammals.

2.5 Polyphenism

Due to their prominent regulatory roles in numerous physiological processes throughout the insect life cycle, changes in ecdysteroid titre can induce highly diverse phenotypes from the same genotype, a phenomenon called polyphenism (Nijhout, 1999, 2003; Simpson et al., 2011). Indeed, ecdysteroids are known to be major regulators of phenotypic plasticity in insects (Hartfelder and Emlen, 2012). This includes caste polyphenism in social insects (Miura, 2004), diapause at different life stages (Denlinger, 2002; Denlinger et al., 2012), and colour and pattern polyphenism in

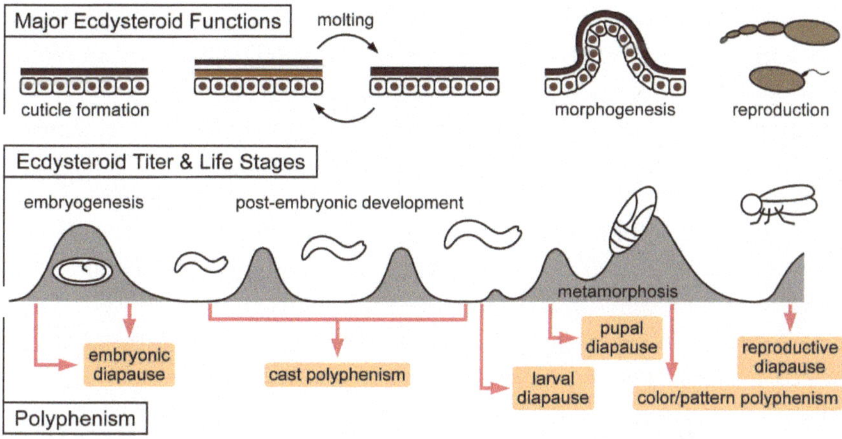

Fig. 2 Major functions of ecdysteroids and various forms of polyphenism caused by their alterations during insect development. Embryonic diapause can occur during both early and late embryogenesis (Denlinger et al., 2012). In Isoptera (termites), larval moulting is co-opted as a process to induce caste differentiation (Hartfelder and Emlen, 2012; Korb and Hartfelder, 2008; Miura, 2004), whereas differential ecdysteroid signalling during development leads to the division of labour among adult females in Hymenoptera (Hartfelder et al., 2000; Strambi et al., 1984; Suzzoni et al., 1980, 1983). Various forms of larval/pupal diapause are essentially the cessation of metamorphosis, and they mostly happen either in the last larval instar or during early adult development (Denlinger et al., 2012). External adult structures often display prominent colour/pattern polyphenism, which is induced by differential pupal ecdysteroid titre (Beldade and Brakefield, 2002; Emlen and Nijhout, 2000; Hartfelder and Emlen, 2012). Differential ecdysteroid signalling also induces reproductive diapause in adult females (Kučerová et al., 2016; Wu et al., 1994).

external adult structures (Beldade and Brakefield, 2002; Emlen and Nijhout, 2000). Fig. 2 summarizes the major functions of ecdysteroids at different life stages in insects, as well as how alterations of these functions can lead to various forms of polyphenism.

3. Six steps of ecdysteroid signalling and their regulation

As described above, physiological functions of ecdysteroids are extremely diverse in different tissues and cell types, at different stages during the life cycle, and among different insect species. Differential environmental conditions can also affect ecdysteroid signalling and induce phenotypic diversity. How can a single hormone signalling pathway mediate such pleiotropic functions? In order to understand how diverse functions of ecdysteroids are implemented at the molecular level, it is critical to identify all the key molecular components and characterize core cellular/molecular processes. Establishment of such core knowledge would allow us to further investigate how such common processes are modified and regulated by different factors under different physiological and developmental contexts.

In order to comprehend the complex series of events that take place during ecdysteroid signalling, here I propose to divide the signalling into the following six steps: biosynthesis, secretion, circulation, transport, modification, and gene expression regulation (Fig. 3). Below I summarize core molecular components involved in each step and describe how their functions are modulated by other factors.

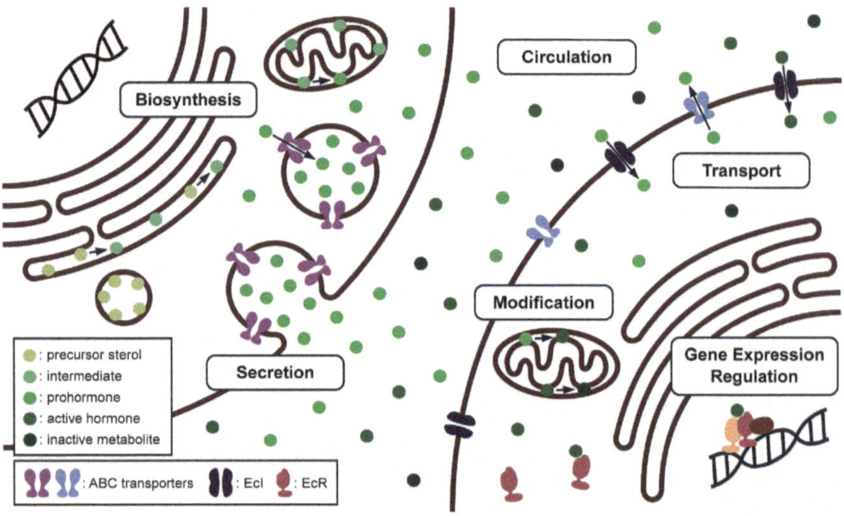

Fig. 3 Six steps of ecdysteroid signalling. Enzymatic reactions involved in prohormone biosynthesis as well as later modifications (activation/inactivation of hormones) happen in membranous cytoplasmic organelles such as the endoplasmic reticulum and mitochondria (Gilbert et al., 2002; Rewitz et al., 2006). ATP-binding cassette (ABC) transporters are involved in active transport of ecdysteroids against the concentration gradient, which happens during prohormone loading into secretory vesicles, as well as hormone elimination out of target cells (Hock et al., 2000; Yamanaka et al., 2015). In contrast, EcI is a solute carrier transporter that incorporates ecdysteroids into target cells down the concentration gradient (Okamoto et al., 2018). EcR is a nuclear receptor that binds to active ecdysteroids and regulates gene expression together with multiple cofactors (Henrich, 2012; Riddiford et al., 2000).

3.1 Biosynthesis

As insects are sterol auxotrophs, sterols that become starting materials for ecdysteroid biosynthesis (or ecdysteroidogenesis) need to be consumed in their diet (Behmer and Nes, 2003; Carvalho et al., 2010; Clark and Bloch, 1959; Clayton, 1964; Lavrynenko et al., 2015). After transport into steroidogenic tissues via lipophorin-mediated trafficking (Igarashi et al., 2018; Talamillo et al., 2013), dietary sterols such as cholesterol or phytosterols go through a series of hydroxylation and oxidation steps mediated by Halloween enzymes (Gilbert, 2004; Gilbert et al., 2002; Iga and Kataoka, 2012; Niwa and Niwa, 2014; Rewitz et al., 2006). It is now established that early steps of ecdysteroidogenesis are mediated by various classes of enzymes, including a [2Fe-2S] Rieske oxygenase named Neverland, cytochrome P450 monooxygenases named Spook and Spookier, and a short-chain dehydrogenase/reductase named Shroud (Namiki et al., 2005; Niwa et al., 2010; Ono et al., 2006; Ou et al., 2011; Yoshiyama et al., 2006; Yoshiyama-Yanagawa et al., 2011). In contrast, subsequent hydroxylation steps are all catalysed by P450 enzymes, namely Phantom, Disembodied, and Shadow (Chávez et al., 2000; Niwa et al., 2004; Warren et al., 2002, 2004). As these enzymes function in membranous organelles such as the endoplasmic reticulum and mitochondria (Gilbert et al., 2002; Rewitz et al., 2006), precursor storage and subcellular trafficking via vesicles can be a critical regulatory step in ecdysteroidogenesis (Chanut-Delalande et al., 2014; Danielsen et al., 2016; Enya et al., 2014, 2015; Ou et al., 2016; Pan et al., 2019; Texada et al., 2019a,b). Although the larval prothoracic gland (PG) is by far the best studied steroidogenic tissue in insects, ecdysteroidogenesis happens in distinct tissues at different life stages, and it is regulated by multiple factors as described below.

3.1.1 Ecdysteroidogenesis during embryonic development

Expression patterns of the Halloween genes suggest that epidermal cells are the major source of ecdysteroids during embryogenesis (Chávez et al., 2000; Gilbert, 2004; Warren et al., 2002). Although the mutant phenotypes mentioned above clearly suggest critical functions of zygotic expression of the Halloween genes, maternally loaded ecdysteroids are also known to have important roles during embryogenesis of many insect species, including the tobacco hornworm *Manduca sexta* and the silkworm *Bombyx mori* (Dinan, 1997; Fujinaga et al., 2020; Hoffmann and Lagueux, 1985; Sonobe and Yamada, 2004; Thompson et al., 1988).

3.1.2 Ecdysteroidogenesis during post-embryonic development

According to the classical scheme established by early insect physiologists, an endocrine factor from the brain, now known as the prothoracicotropic hormone (PTTH), stimulates the production and release of ecdysone from the PG during larval stages (Doane, 1973; Ishizaki and Suzuki, 1994). PTTH activates its receptor Torso in the PG (Rewitz et al., 2009) and upregulates Halloween gene expression for ecdysone production (Namiki et al., 2005; Niwa et al., 2005, 2010; Ou et al., 2016; Yamanaka et al., 2007). Importantly, recent mutant analyses in *B. mori* and *D. melanogaster* demonstrate that PTTH is dispensable for ecdysone production or larval development *per se* (Shimell et al., 2018; Uchibori-Asano et al., 2017). It rather controls the developmental schedule in response to physiological states, such as circadian rhythm and tissue injury (Christensen et al., 2020; Colombani et al., 2015; Deveci et al., 2019; Garelli et al., 2015; Hackney and Cherbas, 2014; Imura et al., 2020; Jaszczak et al., 2016; McBrayer et al., 2007; Mohamed et al., 2014; Selcho et al., 2017; Vafopoulou et al., 2007; Vallejo et al., 2015).

In addition to PTTH, numerous other factors regulate ecdysone production in the PG (Carlisle and Ellis, 1968; Cruz et al., 2020; Di Cara and King-Jones, 2016; Hua et al., 1997, 1999; Liu et al., 2018b; Mala et al., 1977; Mirth et al., 2014; Mizoguchi and Ishizaki, 1982; Ogihara et al., 2015; Ohhara et al., 2015; Richard and Gilbert, 1991; Rodenfels et al., 2014; Sakurai et al., 1989a; Shimada-Niwa and Niwa, 2014; Vafopoulou and Steel, 1991; Yamada et al., 2017; Yamanaka et al., 2005, 2006, 2007, 2010, 2011; Zhang et al., 2018b). Most importantly, insulin/target of rapamycin (TOR) signalling is thoroughly demonstrated in a few model insect species to be a primary factor for regulation of ecdysteroidogenesis in response to nutrients, thus providing a checkpoint to confirm that enough nutrients have been stored to complete adult development (Caldwell et al., 2005; Colombani et al., 2005; Danielsen et al., 2016; Gu et al., 2015; Kemirembe et al., 2012; Koyama et al., 2014; Layalle et al., 2008; Mirth et al., 2005; Ohhara et al., 2017; Pan et al., 2019; Smith et al., 2014; Storelli et al., 2011; Texada et al., 2019a; Walkiewicz and Stern, 2009; Zeng et al., 2020). Interestingly, for some insects where larval food is limited to the provision provided by parents, such as the dung beetle *Onthophagus taurus* and the solitary bee *Osmia lignaria*, exhaustion of the food supply rather than the nutrient checkpoint activates ecdysteroid production in the PG (Helm et al., 2017; Shafiei et al., 2001). The PG thus responds to a variety of signals and controls moulting through ecdysone production in a species-specific manner (Koyama et al., 2020; Marchal et al., 2010; Pan et al., 2020; Texada et al., 2020; Xu et al., 2020a; Yamanaka et al., 2013).

3.1.3 Ecdysteroidogenesis in the adult stage

Reproductive tissues are the major source of ecdysteroids in adult insects. Since ecdysteroidogenesis in the ovary was first reported in mosquitoes (Hagedorn et al., 1975), ecdysteroid production in the gonad of both sexes has been reported in various insect species (Brown et al., 2009). In adult females, ovarian follicle cells produce ecdysteroids (Goltzene et al., 1978; Lagueux et al., 1977; Zhu et al., 1983), whereas ecdysteroid biosynthesis in the male testes remains obscure (Gillott and Ismail, 1995; Koolman et al., 1979). Several lines of evidence rather suggest that the accessory gland in the male reproductive system is the source of ecdysteroids (Hentze et al., 2013; Pondeville et al., 2008).

Ovarian ecdysteroidogenesis is often controlled by brain-derived neuropeptides in species that undergo multiple cycles of reproduction, such as mosquitoes (Brown et al., 1998, 2008; Dhara et al., 2013; Roy et al., 2018; Vogel et al., 2015). Neurohormones with potential ecdysteroidogenic effects on the ovary have also been reported in other insects including migratory locust *Locusta migratoria* and the housefly *Musca domestica* (Adams et al., 1997; Couillaud et al., 1989; Girardie et al., 1991), although their precise functions remain mostly unknown.

3.2 Secretion

In contrast to the extensive studies on ecdysteroidogenesis and its regulation described above, little is known regarding secretory mechanisms of ecdysteroids from steroidogenic tissues into the haemolymph. This is due to the widespread assumption that steroid hormones pass freely through lipid bilayers due to their lipophilic character, a process called simple diffusion. Recent studies in *D. melanogaster*, however, demonstrate that ecdysone produced in the PG is loaded into secretory vesicles and released into the haemolymph through exocytosis (Yamanaka et al., 2015). Upstream signals that induce ecdysone secretion from the PG have not been well examined so far (Cruz et al., 2020; Yamanaka et al., 2015), and cooperative as well as distinct regulation of ecdysteroid production and secretion needs further investigation.

Although the detailed machinery of ecdysone secretion has only been described in *D. melanogaster*, relevant studies involving ultrastructural localization of ecdysone in the PG have been conducted in the waxworm *Galleria mellonella*, which suggest that ecdysone in the PG is likely concentrated into secretory granules that eventually fuse with the plasma membrane (Birkenbeil, 1983, Birkenbeil et al., 1979). Secretory mechanisms of ecdysteroids from other steroidogenic tissues have not been well investigated.

3.3 Circulation

Ecdysteroids travel through the open circulatory system to reach their target tissues. Unlike circulatory transport of mammalian steroid hormones, there appear to be no major ecdysteroid binding proteins essential for transport in the insect circulatory system (Gilbert and Chino, 1974; Spindler et al., 2009), although a few hemolymph proteins that bind to ecdysteroids have been reported (Jaenicke et al., 1999; Whitehead, 1989). There are a few reports on the excretion of ecdysteroids from the circulatory system (Feyereisen et al., 1976; Hoffmann et al., 1974; Robinson et al., 1987; Steel et al., 1982), but the clearance mechanisms of ecdysteroids and their metabolites remain to be investigated.

3.4 Transport

As described below, canonical functions of ecdysteroids are mediated by intracellular nuclear receptors, and ecdysteroids need to enter their target tissues to exert their physiological effects. Due to the pervasive simple diffusion model mentioned above, it has long been assumed that ecdysteroids do not require any membrane proteins to enter the cells. However, a recent study identified a membrane transporter that mediates cellular uptake of ecdysteroids in *D. melanogaster* (Okamoto et al., 2018). In this study, two independent screens identified the same transporter-encoding gene, *Oatp74D*, as an indispensable component of ecdysteroid signalling. *Oatp74D*, renamed as *Ecdysone Importer* (*EcI*), is a member of the solute carrier organic anion transporter gene superfamily that encodes an organic anion transporting polypeptide (OATP) with 12 predicted transmembrane domains. EcI is localized on the cell membrane and is both necessary and sufficient for incorporating 20E into the cells and inducing 20E-dependent gene expression (Okamoto et al., 2018). These results clearly challenge the simple diffusion model of ecdysteroid transport across cell membranes, and instead suggest a

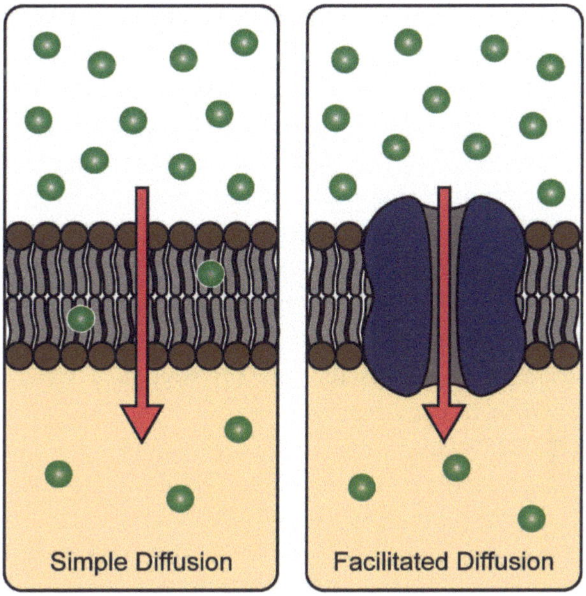

Fig. 4 Models for cellular incorporation of ecdysteroids. It has been long assumed that ecdysteroids (green) enter target cells by simple diffusion (left). Recent work in flies (Okamoto et al., 2018) now indicates that they are incorporated into the cells through a membrane transporter named EcI (blue), supporting a facilitated diffusion model (right).

facilitated diffusion model (Fig. 4). Moreover, EcI is highly expressed in the blood-brain barrier and required for ecdysteroid incorporation into the central nervous system (Okamoto and Yamanaka, 2020). Considering that *EcI* is highly conserved among arthropods that utilize ecdysteroids as the moulting hormone (Okamoto et al., 2018), it is likely that EcI-mediated cellular uptake of ecdysteroids occurs in all arthropod species. Together with the active elimination of ecdysteroids out of the cells (Hock et al., 2000), this novel molecular machinery of steroid hormone transport needs to be investigated further in other species.

3.5 Modification

During post-embryonic development, the PG does not secrete 20-hydroxylated active ecdysteroids such as 20E or makisterone A (Fig. 1); it instead releases prohormones such as ecdysone and 20-deoxymakisterone A (Blais et al., 2010; Gilbert et al., 2002; King et al., 1974; Lavrynenko et al., 2015; Redfern, 1984). In some lepidopteran species, the PG releases

3-dehydroecdysone, which is enzymatically converted into ecdysone in the haemolymph (Chen et al., 1996b, 1999; Kiriishi et al., 1990; Sakurai et al., 1989b). These prohormones are converted into active ecdysteroids in peripheral tissues through 20 carbon hydroxylation mediated by the P450 monooxygenase Shade (Petryk et al., 2003). 20-hydroxylated ecdysteroids are detected as major peaks in the haemolymph, suggesting that they are released back into the haemolymph after synthesis in peripheral tissues such as the gut and fat body (Hikiba et al., 2013; Lavrynenko et al., 2015). Although regulatory mechanisms of Shade expression in non-steroidogenic tissues are of great interest (Buhler et al., 2018), further studies are required to understand the physiological significance of the peripheral activation of ecdysteroids.

Similar to their activation, enzymatic inactivation of ecdysteroids is also conducted in non-steroidogenic tissues, including 26-hydroxylation to 26-oic acids by a P450 monooxygenase CYP18A1 (Guittard et al., 2011; Li et al., 2014; Rewitz et al., 2010; Williams et al., 2000), as well as 3-epimerization by ecdysone oxidase and 3-dehydroecdysone 3α-reductase (Chen et al., 1996a; Li et al., 2015; Sun et al., 2012; Takeuchi et al., 2000, 2001; Yang et al., 2011). Interestingly, expression of some of these enzymes are induced by ecdysteroids, thus forming a negative feedback loop that ensures the transient nature of ecdysteroid signalling necessary for timely and orderly scheduling of cellular events (Davies et al., 2006; Rewitz et al., 2010; Sun et al., 2017).

3.6 Gene expression regulation

Once incorporated into their target tissues, 20E and other bioactive ecdysteroids bind to the nuclear receptor heterodimer composed of the ecdysone receptor (EcR) and Ultraspiracle (USP) to exert their physiological effects (Henrich, 2012; Nakagawa and Henrich, 2009; Riddiford et al., 2000; Thomas et al., 1993; Yao et al., 1992, 1993), although there are a few reported cases where USP is dispensable (Costantino et al., 2008; Sharma et al., 2017). In principle, EcR/USP functions as a ligand-dependent transcription factor that induces expression of a plethora of genes upon binding of active ecdysteroids. As originally proposed by Ashburner et al. to explain ecdysteroid-induced polytene chromosome puffing patterns in *D. melanogaster* (Ashburner, 1990; Ashburner et al., 1974; Hill et al., 2013), the ligand-bound EcR/USP first induces expression of a small set of genes (termed early genes), many of which encode transcription factors.

The early gene products in turn trigger gene expression cascades, which leads to differential expression of multiple downstream genes (Andres et al., 1993; Beckstead et al., 2005; Li and White, 2003; Thummel, 1996). Such differential waves of gene expression are combined with tissue- and stage-specific expression of multiple coactivators and corepressors, thus creating dynamic and diverse functions of ecdysteroids throughout the insect life cycle (Badenhorst et al., 2005; Bai et al., 2000; Gauthier et al., 2012; Kreher et al., 2017; Tsai et al., 1999; Uyehara and McKay, 2019; Uyehara et al., 2017).

EcR/USP-mediated gene expression is under control of multiple external signals. For example, phosphorylation of EcR and USP in response to different signals can affect their dimerization, nuclear trafficking, and DNA binding (Chen et al., 2017; Johnston et al., 2011; Song et al., 2003; Wang et al., 2012). Most importantly, ecdysteroid-induced gene expression is under control of juvenile hormone (JH), another lipophilic hormone with an acyclic sesquiterpenoid structure (Goodman and Cusson, 2012). The major function of JH is to prevent metamorphosis and maintain a juvenile status during development (Jindra et al., 2013, 2015; Riddiford, 2012; Truman, 2019; Truman and Riddiford, 2019). JH exerts its "status quo" action by inducing expression of the transcription factor Krüppel homolog 1 (Kr-h1), a master repressor of adult metamorphosis (Belles and Santos, 2014; Martín et al., 2020). Kr-h1 prevents metamorphosis by repressing expression of the ecdysteroid-inducible transcription factor E93, which works as an adult specifier in both hemimetabolous and holometabolous insects (Chafino et al., 2019; Nystrom et al., 2020; Ureña et al., 2014, 2016). In adult mosquitoes, JH also regulates mRNA splicing and promotes production of specific isoforms of Taiman, a key transcriptional regulator for ecdysteroid-mediated gene expression (Liu et al., 2018a). JH thus modulates the activity of ecdysteroids at different life stages and acts as a stage-specific regulator of ecdysteroid-mediated gene expression in insects.

4. Concluding remarks

When we review each step of ecdysteroid signalling in detail as outlined above, it becomes clear that our current body of knowledge is heavily focused on the first and last steps: biosynthesis and gene expression regulation. Other intermediate steps, namely secretion, circulation, transport, and modification, have not been extensively investigated in any insect species, which creates a huge knowledge gap that prevents our comprehensive understanding of this hormone signalling pathway. In particular, molecular mechanisms of ecdysteroid secretion and cellular uptake have been revealed only recently in *D. melanogaster* larvae (Okamoto et al., 2018; Yamanaka et al., 2015), and further investigation in other life stages as well as in other insect species is clearly warranted. Recent studies have also elucidated non-canonical ecdysteroid signalling pathways, such as those mediated by G protein-coupled receptors (Cai et al., 2014; Evans et al., 2014; Ishimoto et al., 2013; Kang et al., 2019; Manaboon et al., 2009; Srivastava et al., 2005; Wang et al., 2015; Zhao, 2020). Such non-genomic actions of ecdysteroids can modulate the canonical signalling pathway mediated by EcR/USP, but they also have their own independent functions. These studies are expected to expand our knowledge on how ecdysteroids regulate various physiological processes in insects.

Not only does ecdysteroid research contribute to our basic understanding of steroid hormone signalling, but it may also lead to the development of novel insect control agents (Pener and Dhadialla, 2012; Smagghe et al., 2012). Almost a century after Kopeć's initial description of the endocrine control of insect moulting (Kopeć, 1922), we are still only halfway through our journey to understand the complexity and significance of this hormone signalling pathway.

Acknowledgements

Research in the author's laboratory cited here was supported by an NIH Director's New Innovator Award DP2 GM132929, a research grant from the W.M. Keck Foundation, and a Pew Biomedical Scholars Award from the Pew Charitable Trusts.

References

Ables, E.T., Drummond-Barbosa, D., 2010. The steroid hormone ecdysone functions with intrinsic chromatin remodeling factors to control female germline stem cells in Drosophila. Cell Stem Cell 7, 581–592.

Ables, E.T., Hwang, G.H., Finger, D.S., Hinnant, T.D., Drummond-Barbosa, D., 2016. A genetic mosaic screen reveals ecdysone-responsive genes regulating Drosophila oogenesis. G3 (Bethesda) 6, 2629–2642.

Adams, T.S., Dillwith, J.W., Blomquist, G.J., 1984. The role of 20-hydroxyecdysone in housefly sex pheromone biosynthesis. J. Insect Physiol. 30, 287–294.

Adams, T.S., Gerst, J.W., Masler, E.P., 1997. Regulation of ovarian ecdysteroid production in the housefly, Musca domestica. Arch. Insect Biochem. Physiol. 35, 135–148.

Agui, N., 1977. Time studies of ecdysone-action on in vitro apolysis of Chilo suppressalis integument. J. Insect Physiol. 23, 837–842.

Ahmed, S.M.H., Maldera, J.A., Krunic, D., Paiva-Silva, G.O., Pénalva, C., Teleman, A.A., Edgar, B.A., 2020. Fitness trade-offs incurred by ovary-to-gut steroid signalling in Drosophila. Nature 584, 415–419.

Aizenzon, M.G., Zhimulev, I.F., 1975. Hormonal control of lysis of the salivary gland in larvae of Drosophila melanogaster. Inst. Cytol. Genet. 221, 91–93.

Ameku, T., Niwa, R., 2016. Mating-induced increase in germline stem cells via the neuroendocrine system in female Drosophila. PLoS Genet. 12, e1006123.

Andres, A.J., Fletcher, J.C., Karim, F.D., Thummel, C.S., 1993. Molecular analysis of the initiation of insect metamorphosis: a comparative study of Drosophila ecdysone-regulated transcription. Dev. Biol. 160, 388–404.

Ashburner, M., 1990. Puffs, genes, and hormones revisited. Cell 61, 1–3.

Ashburner, M., Chihara, C., Meltzer, P., Richards, G., 1974. Temporal control of puffing activity in polytene chromosomes. Cold Spring Harb. Symp. Quant. Biol. 38, 655–662.

Audit-Lamour, C., Busson, D., 1981. Oogenesis defects in the ecd-1 mutant of Drosophila melanogaster, deficient in ecdysteroid at high temperature. J. Insect Physiol. 27, 829–837.

Awasaki, T., Lee, T., 2011. Orphan nuclear receptors control neuronal remodeling during fly metamorphosis. Nat. Neurosci. 14, 6–7.

Badenhorst, P., Xiao, H., Cherbas, L., Kwon, S.Y., Voas, M., Rebay, I., Cherbas, P., Wu, C., 2005. The Drosophila nucleosome remodeling factor NURF is required for Ecdysteroid signaling and metamorphosis. Genes Dev. 19, 2540–2545.

Bai, J., Uehara, Y., Montell, D.J., 2000. Regulation of invasive cell behavior by taiman, a Drosophila protein related to AIB1, a steroid receptor coactivator amplified in breast cancer. Cell 103, 1047–1058.

Beckstead, R.B., Lam, G., Thummel, C.S., 2005. The genomic response to 20-hydroxyecdysone at the onset of Drosophila metamorphosis. Genome Biol. 6, R99.

Behmer, S.T., Nes, W.D., 2003. Insect sterol nutrition and physiology: a global overview. In: Simpson, S. (Ed.), Advances in Insect Physiology. vol. 31. Academic Press, London, pp. 1–72.

Beldade, P., Brakefield, P.M., 2002. The genetics and evo-devo of butterfly wing patterns. Nat. Rev. Genet. 3, 442–452.

Belles, X., Piulachs, M.D., 2015. Ecdysone signalling and ovarian development in insects: from stem cells to ovarian follicle formation. Biochim. Biophys. Acta 1849, 181–186.

Belles, X., Santos, C.G., 2014. The MEKRE93 (Methoprene tolerant-Krüppel homolog 1-E93) pathway in the regulation of insect metamorphosis, and the homology of the pupal stage. Insect Biochem. Mol. Biol. 52, 60–68.

Bernardi, F., Romani, P., Tzertzinis, G., Gargiulo, G., Cavaliere, V., 2009. EcR-B1 and Usp nuclear hormone receptors regulate expression of the VM32E eggshell gene during Drosophila oogenesis. Dev. Biol. 328, 541–551.

Birkenbeil, H., 1983. Ultrastructural and immunocytochemical investigation of ecdysteroid secretion by the prothoracic gland of the waxmoth Galleria mellonella. Cell Tissue Res. 229, 433–441.

Birkenbeil, H., Eckert, M., Gersch, M., 1979. Electron microscopical-immunocytochemical evidence of ecdysteroids in the prothoracic gland of Galleria mellonella. Cell Tissue Res. 200, 285–290.

Blais, C., Blasco, T., Maria, A., Dauphin-Villemant, C., Lafont, R., 2010. Characterization of ecdysteroids in Drosophila melanogaster by enzyme immunoassay and nano-liquid chromatography–tandem mass spectrometry. J. Chromatogr. B Analyt. Technol. Biomed. Life Sci. 878, 925–932.

Bloch, G., Hefetz, H., Hartfelder, K., 2000. Ecdysteroid titer, ovary status, and dominance in adult worker and queen bumble bees (Bombus terrestris). J. Insect Physiol. 46, 1033–1040.

Blomquist, G.J., Adams, T.S., Dillwith, J.W., 1984. Induction of female sex pheromone production in male houseflies by ovarian implants or 20-hydroxyecdysone. J. Insect Physiol. 30, 295–302.

Blomquist, G.J., Tillman, J.A., Reed, J.R., Gu, P., Vanderwel, D., Choi, S., Reitz, R.C., 1995. Regulation of enzymatic activity involved in sex pheromone production in the housefly, Musca domestica. Insect Biochem. Mol. Biol. 25, 751–757.

Bozzolan, F., Duportets, L., Limousin, D., Wycke, M.A., Demondion, E., François, A., Abrieux, A., Debernard, S., 2015. Synaptotagmin I, a molecular target for steroid hormone signaling controlling the maturation of sexual behavior in an insect. FEBS J. 282, 1432–1444.

Brown, H.L., Truman, J.W., 2009. Fine-tuning of secondary arbor development: the effects of the ecdysone receptor on the adult neuronal lineages of the Drosophila thoracic CNS. Development 136, 3247–3256.

Brown, M.R., Graf, R., Swiderek, K.M., Fendley, D., Stracker, T.H., Champagne, D.E., Lea, A.O., 1998. Identification of a steroidogenic neurohormone in female mosquitoes. J. Biol. Chem. 273, 3967–3971.

Brown, M.R., Clark, K.D., Gulia, M., Zhao, Z., Garczynski, S.F., Crim, J.W., Suderman, R.J., Strand, M.R., 2008. An insulin-like peptide regulates egg maturation and metabolism in the mosquito Aedes aegypti. Proc. Natl. Acad. Sci. U. S. A. 105, 5716–5721.

Brown, M.R., Sieglaff, D.H., Rees, H.H., 2009. Gonadal ecdysteroidogenesis in Arthropoda: occurrence and regulation. Annu. Rev. Entomol. 54, 105–125.

Buhler, K., Clements, J., Winant, M., Bolckmans, L., Vulsteke, V., Callaerts, P., 2018. Growth control through regulation of insulin signalling by nutrition-activated steroid hormone in Drosophila. Development 145, dev165654.

Buszczak, M., Freeman, M.R., Carlson, J.R., Bender, M., Cooley, L., Segraves, W.A., 1999. Ecdysone response genes govern egg chamber development during mid-oogenesis in Drosophila. Development 126, 4581–4589.

Butenandt, A., Karlson, P., 1954. Über die isolierung eines metamorphose-hormons der insekten in kristallisierter form. Z. Naturforsch. 9b, 389–391.

Cai, M.J., Dong, D.J., Wang, Y., Liu, P.C., Liu, W., Wang, J.X., Zhao, X.F., 2014. G-protein-coupled receptor participates in 20-hydroxyecdysone signaling on the plasma membrane. Cell Commun. Signal 12, 9.

Cakouros, D., Daish, T.J., Kumar, S., 2004. Ecdysone receptor directly binds the promoter of the Drosophila caspase dronc, regulating its expression in specific tissues. J. Cell Biol. 165, 631–640.

Caldwell, P., Walkiewicz, M., Stern, M., 2005. Ras activity in the Drosophila prothoracic gland regulates body size and developmental rate via ecdysone release. Curr. Biol. 15, 1785–1795.

Carlisle, D.B., Ellis, P.E., 1968. Hormonal inhibition of the prothoracic gland by the brain in locusts. Nature 220, 706–707.

Carvalho, M., Schwudke, D., Sampaio, J.L., Palm, W., Riezman, I., Dey, G., Gupta, G.D., Mayor, S., Riezman, H., Shevchenko, A., Kurzchalia, T.V., Eaton, S., 2010. Survival strategies of a sterol auxotroph. Development 137, 3675–3685.

Chafino, S., Ureña, E., Casanova, J., Casacuberta, E., Franch-Marro, X., Martín, D., 2019. Upregulation of E93 gene expression acts as the trigger for metamorphosis independently of the threshold size in the beetle Tribolium castaneum. Cell Rep. 27, 1039–1049.e2.

Champlin, D.T., Truman, J.W., 1998. Ecdysteroid control of cell proliferation during optic lobe neurogenesis in the moth Manduca sexta. Development 125, 269–277.

Chanut-Delalande, H., Hashimoto, Y., Pelissier-Monier, A., Spokony, R., Dib, A., Kondo, T., Bohere, J., Niimi, K., Latapie, Y., Inagaki, S., Dubois, L., Valenti, P., Polesello, C., Kobayashi, S., Moussian, B., White, K., Plaza, S., Kageyama, Y., Payre, F., 2014. Pri peptides are mediators of ecdysone for the temporal control of development. Nat. Cell Biol. 16, 1035–1044.

Charles, J.P., 2010. The regulation of expression of insect cuticle protein genes. Insect Biochem. Mol. Biol. 40, 205–213.

Chatani, F., Ohnishi, E., 1976. Effect of ecdysone on the ovarian development of Bombyx silkworm. Dev. Growth Differ. 18, 481–484.

Chávez, V.M., Marqués, G., Delbecque, J.P., Kobayashi, K., Hollingsworth, M., Burr, J., Natzle, J.E., O'Connor, M.B., 2000. The Drosophila disembodied gene controls late embryonic morphogenesis and codes for a cytochrome P450 enzyme that regulates embryonic ecdysone levels. Development 127, 4115–4126.

Chen, J.H., Powls, R., Rees, H.H., 1996a. Purification of the enzymes of ecdysteroid 3-epimerization from the midgut of the cotton leafworm, Spodoptera littoralis. Biochem. Soc. Trans. 24, 436S.

Chen, J.H., Webb, T.J., Powls, R., Rees, H.H., 1996b. Purification and characterization of haemolymph 3-dehydroecdysone 3β-reductase in relation to ecdysteroid biosynthesis in the cotton leafworm Spodoptera littoralis. Eur. J. Biochem. 242, 394–401.

Chen, J.H., Turner, P.C., Rees, H.H., 1999. Molecular cloning and characterization of haemolymph 3-dehydroecdysone 3β-reductase from the cotton leafworm, Spodoptera littoralis: a new member of the third superfamily of oxidoreductases. J. Biol. Chem. 274, 10551–10556.

Chen, C.H., Pan, J., Di, Y.Q., Liu, W., Hou, L., Wang, J.X., Zhao, X.F., 2017. Protein kinase C delta phosphorylates ecdysone receptor B1 to promote gene expression and apoptosis under 20-hydroxyecdysone regulation. Proc. Natl. Acad. Sci. U. S. A. 114, E7121–E7130.

Cho, K.H., Daubnerová, I., Park, Y., Zitnan, D., Adams, M.E., 2014. Secretory competence in a gateway endocrine cell conferred by the nuclear receptor βFTZ-F1 enables stage-specific ecdysone responses throughout development in Drosophila. Dev. Biol. 385, 253–262.

Choi, Y.J., Lee, G., Park, J.H., 2006. Programmed cell death mechanisms of identifiable peptidergic neurons in Drosophila melanogaster. Development 133, 2223–2232.

Christensen, C.F., Koyama, T., Nagy, S., Danielsen, E.T., Texada, M.J., Halberg, K.A., Rewitz, K., 2020. Ecdysone-dependent feedback regulation of prothoracicotropic hormone controls the timing of developmental maturation. Development 147, dev188110.

Clark, A.J., Bloch, K., 1959. The absence of sterol synthesis in insects. J. Biol. Chem. 234, 2578–2582.

Clayton, R.B., 1964. The utilization of sterols by insects. J. Lipid Res. 15, 3–19.

Clever, U., Karlson, P., 1960. Induktion von Puff-Veränderungen in Speicheldrüsenchromosomen von Chironomus tentans durch Ecdyson. Exp. Cell Res. 20, 623–626.

Colombani, J., Bianchini, L., Layalle, S., Pondeville, E., Dauphin-Villemant, C., Antoniewski, C., Carré, C., Noselli, S., Léopold, P., 2005. Antagonistic actions of ecdysone and insulins determine final size in Drosophila. Science 310, 667–670.

Colombani, J., Andersen, D.S., Boulan, L., Boone, E., Romero, N., Virolle, V., Texada, M., Léopold, P., 2015. Drosophila Lgr3 couples organ growth with maturation and ensures developmental stability. Curr. Biol. 20, 2723–2729.

Costantino, B.F., Bricker, D.K., Alexandre, K., Shen, K., Merriam, J.R., Antoniewski, C., Callender, J.L., Henrich, V.C., Presente, A., Andres, A.J., 2008. A novel ecdysone receptor mediates steroid-regulated developmental events during the mid-third instar of Drosophila. PLoS Genet. 4, e1000102.

Couillaud, F., Girardie, A., Girardie, J., 1989. Identification of gonadotropic and antigonadotropic factors from the nervous part of the corpora cardiaca in the African locust. Invertebr. Reprod. Dev. 16, 17–22.

Cruz, J., Martín, D., Franch-Marro, X., 2020. Egfr signaling is a major regulator of ecdysone biosynthesis in the drosophila prothoracic gland. Curr. Biol. 30, 1547–1554.e4.

Dalton, J.E., Lebo, M.S., Sanders, L.E., Sun, F., Arbeitman, M.N., 2009. Ecdysone receptor acts in fruitless-expressing neurons to mediate Drosophila courtship behaviors. Curr. Biol. 19, 1447–1452.

Danielsen, E.T., Moeller, M.E., Yamanaka, N., Ou, Q., Laursen, J.M., Soenderholm, C., Zhuo, R., Phelps, B., Tang, K., Zeng, J., Kondo, S., Nielsen, C.H., Harvald, E.B., Faergeman, N.J., Haley, M.J., O'Connor, K.A., King-Jones, K., O'Connor, M.B., Rewitz, K.F., 2016. A Drosophila genome-wide screen identifies regulators of steroid hormone production and developmental timing. Dev. Cell 37, 558–570.

Davies, L., Williams, D.R., Turner, P.C., Rees, H.H., 2006. Characterization in relation to development of an ecdysteroid agonist-responsive cytochrome P450, CYP18A1, in Lepidoptera. Arch. Biochem. Biophys. 453, 4–12.

Denlinger, D.L., 2002. Regulation of diapause. Annu. Rev. Entomol. 47, 93–122.

Denlinger, D.L., Yocum, G.D., Rinehart, J.P., 2012. Hormonal control of diapause. In: Gilbert, L.I. (Ed.), Insect Endocrinology. Academic Press, London, pp. 430–463.

Denton, D., Shravage, B., Simin, R., Mills, K., Berry, D., Baehrecke, E., Kumar, S., 2009. Autophagy, not apoptosis, is essential for midgut cell death in Drosophila. Curr. Biol. 19, 1741–1746.

Denton, D., Aung-Htut, M.T., Lorensuhewa, N., Nicolson, S., Zhu, W., Mills, K., Cakouros, D., Bergmann, A., Kumar, S., 2013. UTX coordinates steroid hormone-mediated autophagy and cell death. Nat. Commun. 4, 2916.

Deveci, D., Martin, F.A., Leopold, P., Romero, N.M., 2019. AstA signaling functions as an evolutionary conserved mechanism timing juvenile to adult transition. Curr. Biol. 29, 813–822.e4.

Dhara, A., Eum, J.H., Robertson, A., Gulia-Nuss, M., Vogel, K.J., Clark, K.D., Graf, R., Brown, M.R., Strand, M.R., 2013. Ovary ecdysteroidogenic hormone functions independently of the insulin receptor in the yellow fever mosquito, Aedes aegypti. Insect Biochem. Mol. Biol. 43, 1100–1108.

Di Cara, F., King-Jones, K., 2016. The circadian clock is a key driver of steroid hormone production in Drosophila. Curr. Biol. 26, 2469–2477.

Diao, F., Mena, W., Shi, J., Park, D., Diao, F., Taghert, P., Ewer, J., White, B.H., 2016. The splice isoforms of the Drosophila ecdysis triggering hormone receptor have developmentally distinct roles. Genetics 202, 175–189.

Diao, F., Elliott, A.D., Diao, F., Shah, S., White, B.H., 2017. Neuromodulatory connectivity defines the structure of a behavioral neural network. Elife 6, e29797.

Dillard, C., Narbonne-Reveau, K., Foppolo, S., Lanet, E., Maurange, C., 2018. Two distinct mechanisms silence chinmo in Drosophila neuroblasts and neuroepithelial cells to limit their self-renewal. Development 145, dev154534.

Dinan, L., 1997. Ecdysteroids in adults and eggs of the house cricket, Acheta domesticus (Orthoptera: Gryllidae). Comp. Biochem. Physiol. B Biochem. Mol. Biol. 116, 129–135.

Dinan, L., 2009. The Karlson Lecture. Phytoecdysteroids: what use are they? Arch. Insect Biochem. Physiol. 72, 126–141.

Dinan, L., Harmatha, J., Volodin, V., Lafont, R., 2009. Phytoecdysteroids: diversity, biosynthesis and distribution. In: Smagghe, G. (Ed.), Ecdysone: Structures and Functions. Springer Netherlands, Dordrecht, pp. 3–45.

Doane, W.W., 1973. Role of hormones in insect development. In: Counce, S.J., Waddington, C.H. (Eds.), Developmental Systems: Insects. Academic Press, London, pp. 291–497.

Duportets, L., Maria, A., Vitecek, S., Gadenne, C., Debernard, S., 2013. Steroid hormone signaling is involved in the age-dependent behavioral response to sex pheromone in the adult male moth Agrotis ipsilon. Gen. Comp. Endocrinol. 186, 58–66.

Dye, N.A., Popovic, M., Spannl, S., Etournay, R., Kainmüller, D., Ghosh, S., Myers, E.W., Jülicher, F., Eaton, S., 2017. Cell dynamics underlying oriented growth of the Drosophila wing imaginal disc. Development 144, 4406–4421.

Emlen, D.J., Nijhout, H.F., 2000. The development and evolution of exaggerated morphologies in insects. Annu. Rev. Entomol. 45, 661–708.

Enya, S., Ameku, T., Igarashi, F., Iga, M., Kataoka, H., Shinoda, T., Niwa, R., 2014. A Halloween gene noppera-bo encodes a glutathione S-transferase essential for ecdysteroid biosynthesis via regulating the behaviour of cholesterol in Drosophila. Sci. Rep. 4, 6586.

Enya, S., Daimon, T., Igarashi, F., Kataoka, H., Uchibori, M., Sezutsu, H., Shinoda, T., Niwa, R., 2015. The silkworm glutathione S-transferase gene noppera-bo is required for ecdysteroid biosynthesis and larval development. Insect Biochem. Mol. Biol. 61, 1–7.

Evans, R.M., Mangelsdorf, D.J., 2014. Nuclear receptors, RXR, and the big bang. Cell 157, 255–266.

Evans, P.D., Bayliss, A., Reale, V., 2014. GPCR-mediated rapid, non-genomic actions of steroids: comparisons between DmDopEcR and GPER1 (GPR30). Gen. Comp. Endocrinol. 195, 157–163.

Feyereisen, R., Lagueux, M., Hoffmann, J.A., 1976. Dynamics of ecdysone metabolism after ingestion and injection in Locusta migratoria. Gen. Comp. Endocrinol. 29, 319–327.

Flatt, T., Heyland, A., Rus, F., Porpiglia, E., Sherlock, C., Yamamoto, R., Garbuzov, A., Palli, S.R., Tatar, M., Silverman, N., 2008. Hormonal regulation of the humoral innate immune response in Drosophila melanogaster. J. Exp. Biol. 211, 2712–2724.

Flaven-Pouchon, J., Alvarez, J.V., Rojas, C., Ewer, J., 2020. The tanning hormone, bursicon, does not act directly on the epidermis to tan the Drosophila exoskeleton. BMC Biol. 18, 17.

Fujinaga, D., Kohmura, Y., Okamoto, N., Kataoka, H., Mizoguchi, A., 2017. Insulin-like growth factor (IGF)-like peptide and 20-hydroxyecdysone regulate the growth and development of the male genital disk through different mechanisms in the silkmoth, Bombyx mori. Insect Biochem. Mol. Biol. 87, 35–44.

Fujinaga, D., Gu, J., Kawahara, H., Ogihara, M.H., Kojima, I., Takeshima, M., Kataoka, H., 2020. Twenty-hydroxyecdysone produced by dephosphorylation and ecdysteroidogenesis regulates early embryonic development in the silkmoth, Bombyx mori. Insect Biochem. Mol. Biol. 127, 103491.

Fukuda, S., 1940. Hormonal control of molting and pupation in the silkworm. Proc. Imp. Acad. (Tokyo) 16, 417–420.

Gancz, D., Lengil, T., Gilboa, L., 2011. Coordinated regulation of niche and stem cell precursors by hormonal signaling. PLoS Biol. 9, e1001202.

Ganter, G.K., Walton, K.L., Merriman, J.O., Salmon, M.V., Brooks, K.M., Maddula, S., Kravitz, E.A., 2007. Increased male–male courtship in ecdysone receptor deficient adult flies. Behav. Genet. 37, 507–512.

Ganter, G.K., Panaitiu, A.E., Desilets, J.B., Davis-Heim, J.A., Fisher, E.A., Tan, L.C., Heinrich, R., Buchanan, E.B., Brooks, K.M., Kenney, M.T., Verde, M.G., Downey, J., Adams, A.M., Grenier, J.S., Maddula, S., Shah, P., Kincaid, K.M., O'Brien, J.R., 2011. Drosophila male courtship behavior is modulated by ecdysteroids. J. Insect Physiol. 57, 1179–1184.

Ganter, G.K., Desilets, J.B., Davis-Knowlton, J.A., Panaitiu, A.E., Sweezy, M., Sungail, J., Tan, L.C., Adams, A.M., Fisher, E.A., O'Brien, J.R., Kincaid, K.M., Heinrich, R., 2012. Drosophila female precopulatory behavior is modulated by ecdysteroids. J. Insect Physiol. 58, 413–419.

Garelli, A., Heredia, F., Casimiro, A.P., Macedo, A., Nunes, C., Garcez, M., Dias, A.R.M., Volonte, Y.A., Uhlmann, T., Caparros, E., 2015. Dilp8 requires the neuronal relaxin receptor Lgr3 to couple growth to developmental timing. Nat. Commun. 6, 8732.

Gauthier, S.A., VanHaaften, E., Cherbas, L., Cherbas, P., Hewes, R.S., 2012. Cryptocephal, the Drosophila melanogaster ATF4, is a specific coactivator for ecdysone receptor isoform B2. PLoS Genet. 8, e1002883.

Geva, S., Hartfelder, K., Bloch, G., 2005. Reproductive division of labor, dominance, and ecdysteroid levels in hemolymph and ovary of the bumble bee Bombus terrestris. J. Insect Physiol. 51, 811–823.

Giesen, K., Lammel, U., Langehans, D., Krukkert, K., Bunse, I., Klambt, C., 2003. Regulation of glial cell number and differentiation by ecdysone and Fos signaling. Mech. Dev. 120, 401–413.

Gilbert, L.I., 2004. Halloween genes encode P450 enzymes that mediate steroid hormone biosynthesis in Drosophila melanogaster. Mol. Cell. Endocrinol. 215, 1–10.

Gilbert, L.I., Chino, H., 1974. Transport of lipids in insects. J. Lipid Res. 15, 439–456.

Gilbert, L.I., Rybczynski, R., Warren, J.T., 2002. Control and biochemical nature of the ecdysteroidogenic pathway. Annu. Rev. Entomol. 47, 883–916.

Gillott, C., Ismail, P.M., 1995. In vitro synthesis of ecdysteroid by the male accessory reproductive glands, testis and abdominal integument of the adult migratory grasshopper, Melanoplus sanguinipes. Invertebr. Reprod. Dev. 27, 65–71.

Girardie, J., Richard, O., Huet, J.C., Nespoulous, C., Van Dorsselaer, A., Girardie, A., 1991. Physical characterization and sequence identification of the ovary maturating parsin. A new neurohormone purified from the nervous corpora cardiaca of the African locust (Locusta migratoria migratoroides). Eur. J. Biochem. 202, 1121–1126.

Goltzene, F., Lagueux, M., Charlet, M., Hoffmann, J.A., 1978. The follicle cell epithelium of maturing ovaries of Locusta migratoria: a new biosynthetic tissue for ecdysone. Hoppe Seylers Z. Physiol. Chem. 359, 1427–1434.

Goodman, W.G., Cusson, M., 2012. The juvenile hormones. In: Gilbert, L.I. (Ed.), Insect Endocrinology. Academic Press, London, pp. 310–365.

Goodwin, T.W., Horn, D.H.S., Karlson, P., Koolman, J., Nakanishi, K., Robbins, W.E., Siddall, J.B., Takemoto, T., 1978. Ecdysteroids: a new generic term. Nature 272, 111.

Gorski, S.M., Chittaranjan, S., Pleasance, E.D., Freeman, J.D., Anderson, C.L., Varhol, R.J., Coughlin, S.M., Zuyderduyn, S.D., Jones, S.J.M., Marra, M.A., 2003. A SAGE approach to discovery of genes involved in autophagic cell death. Curr. Biol. 13, 358–363.

Gruntenko, N.E., Karpova, E.K., Adonyeva, N.V., Chentsova, N.A., Faddeeva, N.V., Alekseev, A.A., Rauschenbach, I.Y., 2005. Juvenile hormone, 20-hydroxyecdysone and dopamine interaction in Drosophila virilis reproduction under normal and nutritional stress conditions. J. Insect Physiol. 51, 417–425.

Gu, S.H., Chen, C.H., Hsieh, Y.C., Lin, P.L., Young, S.C., 2015. Modulatory effects of bombyxin on ecdysteroidogenesis in Bombyx mori prothoracic glands. J. Insect Physiol. 72, 61–69.

Guittard, E., Blais, C., Maria, A., Parvy, J.P., Pasricha, S., Lumb, C., Lafont, R., Daborn, P.J., Dauphin-Villemant, C., 2011. CYP18A1, a key enzyme of Drosophila steroid hormone inactivation, is essential for metamorphosis. Dev. Biol. 349, 35–45.

Hackney, J.F., Cherbas, P., 2014. Injury response checkpoint and developmental timing in insects. Fly (Austin) 8, 226–231.

Hackney, J.F., Pucci, C., Naes, E., Dobens, L., 2007. Ras signaling modulates activity of the ecdysone receptor EcR during cell migration in the Drosophila ovary. Dev. Dyn. 236, 1213–1226.

Hadjieconomou, D., King, G., Gaspar, P., Mineo, A., Blackie, L., Ameku, T., Studd, C., de Mendoza, A., Diao, F., White, B.H., Brown, A.E.X., Plaçais, P.Y., Préat, T., Miguel-Aliaga, I., 2020. Enteric neurons increase maternal food intake during reproduction. Nature 587, 455–459.

Hagedorn, H.H., 2005. The role of ecdysteroids in reproduction. In: Gilbert, L.I., Iatrou, S.S., Gill, S.S. (Eds.), Comprehensive Molecular Insect Science. vol. 7. Elsevier, Oxford, pp. 205–261.

Hagedorn, H.H., Fallon, A.M., 1973. Ovarian control of vitellogenin synthesis by the fat body in Aedes aegypti. Nature 244, 103–105.

Hagedorn, H.H., O'Connor, J.D., Fuchs, M.S., Sage, B., Schlaeger, D.A., Bohm, M.K., 1975. The ovary as a source of alpha-ecdysone in an adult mosquito. Proc. Natl. Acad. Sci. U. S. A. 72, 3255–3259.

Hampshire, F., Horn, D.H.S., 1966. Structure of crustecdysone, a crustacean moulting hormone. Chem. Commun. (Camb) 2, 37–38.

Han, P., Han, J., Fan, J., Zhang, M., Ma, E., Li, S., Fan, R., Zhang, J., 2017. 20-Hydroxyecdysone activates PGRP-SA mediated immune response in Locusta migratoria. Dev. Comp. Immunol. 72, 128–139.

Hara, Y., Hirai, K., Togane, Y., Akagawa, H., Iwabuchi, K., Tsujimura, H., 2013. Ecdysone-dependent and ecdysone-independent programmed cell death in the developing optic lobe of Drosophila. Dev. Biol. 374, 127–141.

Hartfelder, K., Emlen, D.J., 2012. Endocrine control of insect polyphenism. In: Gilbert, L.I. (Ed.), Insect Endocrinology. Academic Press, London, pp. 464–522.

Hartfelder, K., Cnaani, J., Hefetz, A., 2000. Caste-specific differences in ecdysteroid titers in early larval stages of the bumblebee Bombus terrestris. J. Insect Physiol. 46, 1433–1439.

Helm, B.R., Rinehart, J.P., Yocum, G.D., Greenlee, K.J., Bowsher, J.H., 2017. Metamorphosis is induced by food absence rather than a critical weight in the solitary bee, Osmia lignaria. Proc. Natl. Acad. Sci. U. S. A. 114, 10924–10929.

Henrich, V.C., 2012. The ecdysteroid receptor. In: Gilbert, L.I. (Ed.), Insect Endocrinology. Academic Press, London, pp. 177–218.

Hentze, J.L., Moeller, M.E., Jørgensen, A.F., Bengtsson, M.S., Bordoy, A.M., Warren, J.T., Gilbert, L.I., Andersen, O., Rewitz, K.F., 2013. Accessory gland as a site for prothoracicotropic hormone controlled ecdysone synthesis in adult male insects. PLoS One 8, e55131.

Herboso, L., Oliveira, M.M., Talamillo, A., Pérez, C., González, M., Martín, D., Sutherland, J.D., Shingleton, A.W., Mirth, C.K., Barrio, R., 2015. Ecdysone promotes growth of imaginal discs through the regulation of Thor in D. melanogaster. Sci. Rep. 5, 12383.

Hikiba, J., Ogihara, M.H., Iga, M., Saito, K., Fujimoto, Y., Suzuki, M., Kataoka, H., 2013. Simultaneous quantification of individual intermediate steroids in silkworm ecdysone biosynthesis by liquid chromatography–tandem mass spectrometry with multiple reaction monitoring. J. Chromatogr. B Analyt. Technol. Biomed. Life Sci. 915–916, 52–56.

Hill, R.J., Billas, I.M.L., Bonneton, F., Graham, L.D., Lawrence, M.C., 2013. Ecdysone receptors: from the Ashburner model to structural biology. Annu. Rev. Entomol. 58, 251–271.

Hirashima, A., Rauschenbach, I.Y., Sukhanova, M.J., 2000. Ecdysteroids in stress responsive and nonresponsive Drosophila virilis lines under stress conditions. Biosci. Biotechnol. Biochem. 64, 2657–2662.

Hiruma, K., Riddiford, L.M., 2009. The molecular mechanisms of cuticular melanization: the ecdysone cascade leading to dopa decarboxylase expression in Manduca sexta. Insect Biochem. Mol. Biol. 39, 245–253.

Hock, T., Cottrill, T., Keegan, J., Garza, D., 2000. The E23 early gene of Drosophila encodes an ecdysone-inducible ATP-binding cassette transporter capable of repressing ecdysone-mediated gene activation. Proc. Natl. Acad. Sci. U. S. A. 97, 9519–9524.

Hodin, J., Riddiford, L.M., 1998. The ecdysone receptor and ultraspiracle regulate the timing and progression of ovarian morphogenesis during Drosophila metamorphosis. Dev. Genes Evol. 208, 304–317.

Hoffmann, J.A., Lagueux, M., 1985. Endocrine aspects of embryonic development in insects. In: Kerkut, G.A., Gilbert, L.I. (Eds.), Comprehensive Insect Physiology, Biochemistry and Pharmacology. vol. 1. Pergamon Press, Oxford, pp. 435–460.

Hoffmann, J.A., Koolman, J., Karlson, P., Joly, P., 1974. Molting hormone titer and metabolic fate of injected ecdysone during the fifth larval instar and in adults of Locusta migratoria (Orthoptera). Gen. Comp. Endocrinol. 22, 90–97.

Hua, Y.J., Jiang, R.J., Koolman, J., 1997. Multiple control of ecdysone biosynthesis in blowfly larvae: interaction of ecdysiotropins and ecdysiostatins. Arch. Insect Biochem. Physiol. 35, 125–134.

Hua, Y.J., Tanaka, Y., Nakamura, K., Sakakibara, M., Nagata, S., Kataoka, H., 1999. Identification of a prothoracicostatic peptide in the larval brain of the silkworm, Bombyx mori. J. Biol. Chem. 274, 31169–31173.

Hult, E.F., Huang, J., Marchal, E., Lam, J., Tobe, S.S., 2015. RXR/USP and EcR are critical for the regulation of reproduction and the control of JH biosynthesis in Diploptera punctata. J. Insect Physiol. 80, 48–60.

Iga, M., Kataoka, H., 2012. Recent studies on insect hormone metabolic pathways mediated by cytochrome P450 enzymes. Biol. Pharm. Bull. 35, 838–843.

Igarashi, F., Ogihara, M.H., Iga, M., Kataoka, H., 2018. Cholesterol internalization and metabolism in insect prothoracic gland, a steroidogenic organ, via lipoproteins. Steroids 134, 110–116.

Imura, E., Shimada-Niwa, Y., Nishimura, T., Hückesfeld, S., Schlegel, P., Ohhara, Y., Kondo, S., Tanimoto, H., Cardona, A., Pankratz, M.J., Niwa, R., 2020. The Corazonin-PTTH neuronal axis controls systemic body growth by regulating basal ecdysteroid biosynthesis in Drosophila melanogaster. Curr. Biol. 30, 2156–2165.e5.

Ishimoto, H., Kitamoto, T., 2010. The steroid molting hormone Ecdysone regulates sleep in adult Drosophila melanogaster. Genetics 185, 269–281.

Ishimoto, H., Sakai, T., Kitamoto, T., 2009. Ecdysone signaling regulates the formation of long-term courtship memory in adult Drosophila melanogaster. Proc. Natl. Acad. Sci. U. S. A. 106, 6381–6386.

Ishimoto, H., Wang, Z., Rao, Y., Wu, C.F., Kitamoto, T., 2013. A novel role for ecdysone in Drosophila conditioned behavior: linking GPCR-mediated non-canonical steroid action to cAMP signaling in adult brain. PLoS Genet. 9, e1003843.

Ishizaki, H., Suzuki, A., 1994. The brain secretory peptides that control moulting and metamorphosis of the silkmoth, Bombyx mori. Int. J. Dev. Biol. 38, 301–310.

Jaenicke, E., Föll, R., Decker, H., 1999. Spider hemocyanin binds ecdysone and 20-OH-ecdysone. J. Biol. Chem. 274, 34267–34271.

Jaszczak, J.S., Wolpe, J.B., Bhandari, R., Jaszczak, R.G., Halme, A., 2016. Growth coordination during Drosophila melanogaster imaginal disc regeneration is mediated by signaling through the relaxin receptor Lgr3 in the prothoracic gland. Genetics 204, 703–709.

Jiang, C., Baehrecke, E.H., Thummel, C.S., 1997. Steroid regulated programmed cell death during Drosophila metamorphosis. Development 124, 4673–4683.

Jiang, C., Lamblin, A.F., Steller, H., Thummel, C.S., 2000. A steroid-triggered transcriptional hierarchy controls salivary gland cell death during Drosophila metamorphosis. Mol. Cell 5, 445–455.

Jindra, M., Palli, S.R., Riddiford, L.M., 2013. The juvenile hormone signaling pathway in insect development. Annu. Rev. Entomol. 58, 181–204.

Jindra, M., Bellés, X., Shinoda, T., 2015. Molecular basis of juvenile hormone signaling. Curr. Opin. Insect. Sci. 11, 39–46.

Johnston, D.M., Sedkov, Y., Petruk, S., Riley, K.M., Fujioka, M., Jaynes, J.B., Mazo, A., 2011. Ecdysone- and NO-mediated gene regulation by competing EcR/Usp and E75A nuclear receptors during Drosophila development. Mol. Cell 44, 51–61.

Jürgens, G., Kluding, H., Nüsslein-Volhard, C., Wieschaus, E., 1984. Mutations affecting the pattern of the larval cuticle in Drosophila melanogaster. II. Zygotic loci on the third chromosome. Wilhelm Roux Arch. Dev. Biol. 193, 283–295.

Kang, Y., Bashirullah, A., 2014. A steroid-controlled global switch in sensitivity to apoptosis during Drosophila development. Dev. Biol. 386, 34–41.

Kang, X.L., Zhang, J.Y., Wang, D., Zhao, Y.M., Han, X.L., Wang, J.X., Zhao, X.F., 2019. The steroid hormone 20-hydroxyecdysone binds to dopamine receptor to repress lepidopteran insect feeding and promote pupation. PLoS Genet. 15, e1008331.

Kato, Y., Riddiford, L.M., 1987. The role of 20-hydroxyecdysone in stimulating epidermal mitoses during the larval-pupal transformation of the tobacco hornworm, Manduca sexta. Development 100, 227–236.

Kawasaki, H., 1995. Ecdysteroid concentration inducing cell proliferation brings about the imaginal differentiation in the wing disc of Bombyx mori in vitro. Dev. Growth Differ. 37, 575–580.

Kemirembe, K., Liebmann, K., Bootes, A., Smith, W.A., Suzuki, Y., 2012. Amino acids and TOR signaling promote prothoracic gland growth and the initiation of larval molts in the tobacco hornworm Manduca sexta. PLoS One 7, e44429.

Kim, Y.J., Žitňan, D., Cho, K.H., Schooley, D.A., Mizoguchi, A., Adams, M.E., 2006a. Central peptidergic ensembles associated with organization of an innate behavior. Proc. Natl. Acad. Sci. U. S. A. 103, 14211–14216.

Kim, Y.J., Žitňan, D., Galizia, C.G., Cho, K.H., Adams, M.E., 2006b. A command chemical triggers an innate behavior by sequential activation of multiple peptidergic ensembles. Curr. Biol. 16, 1395–1407.

King, D.S., Bollenbacher, W.E., Borst, D.W., Vedeckis, W.V., O'Connor, J.D., Ittycheriah, P.I., Gilbert, L.I., 1974. The secretion of alpha-ecdysone by the prothoracic glands of Manduca sexta in vitro. Proc. Natl. Acad. Sci. U. S. A. 71, 793–796.

Kingan, T.G., Adams, M.E., 2000. Ecdysteroids regulate secretory competence in Inka cells. J. Exp. Biol. 203, 3011–3018.

King-Jones, K., Thummel, C.S., 2005. Nuclear receptors—a perspective from Drosophila. Nat. Rev. Genet. 6, 311–323.

Kiriishi, S., Rountree, D.B., Sakurai, S., Gilbert, L.I., 1990. Prothoracic gland synthesis of 3-dehydroecdysone and its hemolymph 3β-reductase mediated conversion to ecdysone in representative insects. Experientia 46, 716–721.

Knapp, E., Sun, J., 2017. Steroid signaling in mature follicles is important for Drosophila ovulation. Proc. Natl. Acad. Sci. U. S. A. 114, 699–704.

König, A., Yatsenko, A.S., Weiss, M., Shcherbata, H.R., 2011. Ecdysteroids affect Drosophila ovarian stem cell niche formation and early germline differentiation. EMBO J. 30, 1549–1562.

Koolman, J., Scheller, K., Bodenstein, D., 1979. Ecdysteroids in the adult male blowfly, Caliphora vicina. Experientia 35, 134–135.

Kopeć, S., 1922. Studies on the necessity of the brain for the inception of insect metamorphosis. Biol. Bull. 42, 323–342.

Korb, J., Hartfelder, K., 2008. Life history and development: a framework for understanding developmental plasticity in lower termites. Biol. Rev. 83, 295–313.

Koyama, T., Iwami, M., Sakurai, S., 2004. Ecdysteroid control of cell cycle and cellular commitment in insect wing imaginal discs. Mol. Cell. Endocrinol. 213, 155–166.

Koyama, T., Rodrigues, M.A., Athanasiadis, A., Shingleton, A.W., Mirth, C.K., 2014. Nutritional control of body size through FoxO-Ultraspiracle mediated ecdysone biosynthesis. Elife 3, e03091.

Koyama, T., Texada, M.J., Halberg, K.A., Rewitz, K., 2020. Metabolism and growth adaptation to environmental conditions in Drosophila. Cell. Mol. Life Sci. 77, 4523–4551.

Kreher, J., Kovač, K., Bouazoune, K., Mačinković, I., Ernst, A.L., Engelen, E., Pahl, R., Finkernagel, F., Murawska, M., Ullah, I., Brehm, A., 2017. EcR recruits dMi-2 and increases efficiency of dMi-2-mediated remodelling to constrain transcription of hormone-regulated genes. Nat. Commun. 8, 14806.

Kučerová, L., Kubrak, O.I., Bengtsson, J.M., Strnad, H., Nylin, S., Theopold, U., Nässel, D.R., 2016. Slowed aging during reproductive dormancy is reflected in genome-wide transcriptome changes in Drosophila melanogaster. BMC Genomics 17, 50.

Kuo, C.T., Zhu, S., Younger, S., Jan, L.Y., Jan, Y.N., 2006. Identification of E2/E3 ubiquitinating enzymes and caspase activity regulating Drosophila sensory neuron dendrite pruning. Neuron 51, 283–290.

Lafont, R., Dauphin-Villemant, C., Warren, J.T., Rees, H., 2012. Ecdysteroid chemistry and biochemistry. In: Gilbert, L.I. (Ed.), Insect Endocrinology. Academic Press, London, pp. 106–176.

Lagueux, M., Hirn, M., Hoffmann, J.A., 1977. Ecdysone during ovarian development in Locusta migratoria. J. Insect Physiol. 23, 109–119.

Lanet, E., Gould, A.P., Maurange, C., 2013. Protection of neuronal diversity at the expense of neuronal numbers during nutrient restriction in the Drosophila visual system. Cell Rep. 3, 587–594.

Lavrynenko, O., Rodenfels, J., Carvalho, M., Dye, N.A., Lafont, R., Eaton, S., Shevchenko, A., 2015. The ecdysteroidome of Drosophila: influence of diet and development. Development 142, 3758–3768.

Layalle, S., Arquier, N., Léopold, P., 2008. The TOR pathway couples nutrition and developmental timing in Drosophila. Dev. Cell 15, 568–577.

Lee, C.Y., Baehrecke, E.H., 2001. Steroid regulation of autophagic programmed cell death during development. Development 128, 1443–1455.

Lee, T., Marticke, S., Sung, C., Robinow, S., Luo, L., 2000. Cell-autonomous requirement of the USP/EcR-B ecdysone receptor for mushroom body neuronal remodeling in Drosophila. Neuron 28, 807–818.

Lee, C.Y., Cooksey, B.A.K., Baehrecke, E.H., 2002a. Steroid regulation of midgut cell death during Drosophila development. Dev. Biol. 250, 101–111.

Lee, C.Y., Simon, C.R., Woodard, C.T., Baehrecke, E.H., 2002b. Genetic mechanism for the stage- and tissue-specific regulation of steroid triggered programmed cell death in Drosophila. Dev. Biol. 252, 138–148.

Lee, C.Y., Clough, E.A., Yellon, P., Teslovich, T.M., Stephan, D.A., Baehrecke, E.H., 2003. Genome-wide analyses of steroid- and radiation-triggered programmed cell death in Drosophila. Curr. Biol. 13, 350–357.

Li, T.R., White, K.P., 2003. Tissue-specific gene expression and ecdysone-regulated genomic networks in Drosophila. Dev. Cell 5, 59–72.

Li, Z., Ge, X., Ling, L., Zeng, B., Xu, J., Aslam, A.F.M., You, L., Palli, S.R., Huang, Y., Tan, A., 2014. CYP18A1 regulates tissue-specific steroid hormone inactivation in Bombyx mori. Insect Biochem. Mol. Biol. 54, 33–41.

Li, Z., You, L., Zeng, B., Ling, L., Xu, J., Chen, X., Zhang, Z., Palli, S.R., Huang, Y., Tan, A., 2015. Ectopic expression of ecdysone oxidase impairs tissue degeneration in Bombyx mori. Proc. Biol. Sci. 282, 20150513.

Liu, P., Fu, X., Zhu, J., 2018a. Juvenile hormone-regulated alternative splicing of the taiman gene primes the ecdysteroid response in adult mosquitoes. Proc. Natl. Acad. Sci. U. S. A. 115, E7738–E7747.

Liu, S., Li, K., Gao, Y., Liu, X., Chen, W., Ge, W., Feng, Q., Palli, S.R., Li, S., 2018b. Antagonistic actions of juvenile hormone and 20-hydroxyecdysone within the ring gland determine developmental transitions in Drosophila. Proc. Natl. Acad. Sci. U. S. A. 115, 139–144.

Mala, J., Granger, N.A., Sehnal, F., 1977. Control of prothoracic gland activity in larvae of Galleria mellonella. J. Insect Physiol. 23, 309–316.

Manaboon, M., Iga, M., Iwami, M., Sakurai, S., 2009. Intracellular mobilization of Ca2 + by the insect steroid hormone 20-hydroxyecdysone during programmed cell death in silkworm anterior silk glands. J. Insect Physiol. 55, 122–128.

Manning, L., Sheth, J., Bridges, S., Saadin, A., Odinammadu, K., Andrew, D., Spencer, S., Montell, D., Starz-Gaiano, M., 2017. A hormonal cue promotes timely follicle cell migration by modulating transcription profiles. Mech. Dev. 148, 56–68.

Marchal, E., Vandersmissen, H.P., Badisco, L., Van de Velde, S., Verlinden, H., Iga, M., Van Wielendaele, P., Huybrechts, R., Simonet, G., Smagghe, G., Vanden Broeck, J., 2010. Control of ecdysteroidogenesis in prothoracic glands of insects: a review. Peptides 31, 506–519.

Marchetti, G., Tavosanis, G., 2019. Modulators of hormonal response regulate temporal fate specification in the Drosophila brain. PLoS Genet. 15, e1008491.

Martín, D., Chafino, S., Franch-Marro, X., 2020. How stage identity is established in insects: the role of the Metamorphic Gene Network. Curr. Opin. Insect. Sci. 43, 29–38.

McBrayer, Z., Ono, H., Shimell, M., Parvy, J.P., Beckstead, R.B., Warren, J.T., Thummel, C.S., Dauphin-Villemant, C., Gilbert, L.I., O'Connor, M.B., 2007. Prothoracicotropic hormone regulates developmental timing and body size in Drosophila. Dev. Cell 13, 857–871.

Meiselman, M.R., Kingan, T.G., Adams, M.E., 2018. Stress-induced reproductive arrest in Drosophila occurs through ETH deficiency-mediated suppression of oogenesis and ovulation. BMC Biol. 16, 18.

Mena, W., Diegelmann, S., Wegener, C., Ewer, J., 2016. Stereotyped responses of Drosophila peptidergic neuronal ensemble depend on downstream neuromodulators. Elife 5, e19686.

Mendes, C.M., Mirth, C.K., 2016. Stage-specific plasticity in ovary size is regulated by insulin/insulin-like growth factor and ecdysone signalling in Drosophila. Genetics 202, 703–719.

Mirth, C., Truman, J.W., Riddiford, L.M., 2005. The role of the prothoracic gland in determining critical weight for metamorphosis in Drosophila melanogaster. Curr. Biol. 15, 1796–1807.

Mirth, C.K., Truman, J.W., Riddiford, L.M., 2009. The Ecdysone receptor controls the post-critical weight switch to nutrition-independent differentiation in Drosophila wing imaginal discs. Development 136, 2345–2353.

Mirth, C.K., Tang, H.Y., Makohon-Moore, S.C., Salhadar, S., Gokhale, R.H., Warner, R.D., Koyama, T., Riddiford, L.M., Shingleton, A.W., 2014. Juvenile hormone regulates body size and perturbs insulin signaling in Drosophila. Proc. Natl. Acad. Sci. U. S. A. 111, 7018–7023.

Miura, T., 2004. Proximate mechanisms and evolution of caste polyphenism in social insects: from society to genes. Ecol. Res. 19, 141–148.

Mizoguchi, A., Ishizaki, H., 1982. Prothoracic glands of the saturniid moth Samia cynthia ricini possess a circadian clock controlling gut purge timing. Proc. Natl. Acad. Sci. U. S. A. 79, 2726–2730.

Mohamed, A.A.M., Wang, Q., Bembenek, J., Ichihara, N., Hiragaki, S., Suzuki, T., Takeda, M., 2014. N-acetyltransferase (nat) is a critical conjunct of photoperiodism between the circadian system and endocrine Axis in Antheraea pernyi. PLoS One 9, e92680.

Nakagawa, Y., Henrich, V.C., 2009. Arthropod nuclear receptors and their role in molting. FEBS J. 276, 6128–6157.

Namiki, T., Niwa, R., Sakudoh, T., Shirai, K.I., Takeuchi, H., Kataoka, H., 2005. Cytochrome P450 CYP307A1/Spook: a regulator for ecdysone synthesis in insects. Biochem. Biophys. Res. Commun. 337, 367–374.

Nicolson, S., Denton, D., Kumar, S., 2015. Ecdysone-mediated programmed cell death in Drosophila. Int. J. Dev. Biol. 59, 23–32.

Nijhout, H.F., 1999. Control mechanisms of polyphenic development in insects. Bioscience 49, 181–192.

Nijhout, H.F., 2003. Development and evolution of adaptive polyphenisms. Evol. Dev. 5, 9–18.

Nijhout, H.F., Grunert, L.W., 2002. Bombyxin is a growth factor for wing imaginal disks in Lepidoptera. Proc. Natl. Acad. Sci. U. S. A. 99, 15446–15450.

Nijhout, H.F., Smith, W.A., Schachar, I., Subramanian, S., Tobler, A., Grunert, L.W., 2007. The control of growth and differentiation of the wing imaginal disks of Manduca sexta. Dev. Biol. 302, 569–576.

Nijhout, H.F., Laub, E., Grunert, L.W., 2018. Hormonal control of growth in the wing imaginal disks of Junonia coenia: the relative contributions of insulin and ecdysone. Development 145, dev160101.

Ninov, N., Manjón, C., Martín-Blanco, E., 2009. Dynamic control of cell cycle and growth coupling by ecdysone, EGFR, and PI3K signaling in Drosophila histoblasts. PLoS Biol. 7, e1000079.

Nishimura, T., 2020. Feedforward regulation of glucose metabolism by steroid hormones drives a developmental transition in Drosophila. Curr. Biol. 30, 3624–3632.e5.

Niwa, R., Niwa, Y.S., 2014. Enzymes for ecdysteroid biosynthesis: their biological functions in insects and beyond. Biosci. Biotechnol. Biochem. 78, 1283–1292.

Niwa, R., Matsuda, T., Yoshiyama, T., Namiki, T., Mita, K., Fujimoto, Y., Kataoka, H., 2004. CYP306A1, a cytochrome P450 enzyme, is essential for ecdysteroid biosynthesis in the prothoracic glands of Bombyx and Drosophila. J. Biol. Chem. 279, 35942–35949.

Niwa, R., Sakudoh, T., Namiki, T., Saida, K., Fujimoto, Y., Kataoka, H., 2005. The ecdysteroidogenic P450 Cyp302a1/disembodied from the silkworm, Bombyx mori, is transcriptionally regulated by prothoracicotropic hormone. Insect Mol. Biol. 14, 563–571.

Niwa, R., Namiki, T., Ito, K., Shimada-Niwa, Y., Kiuchi, M., Kawaoka, S., Kayukawa, T., Banno, Y., Fujimoto, Y., Shigenobu, S., Kobayashi, S., Shimada, T., Katsuma, S., Shinoda, T., 2010. Non-molting glossy/shroud encodes a short-chain dehydrogenase/reductase that functions in the "Black Box" of the ecdysteroid biosynthesis pathway. Development 137, 1991–1999.

Nüsslein-Volhard, C., Wieschaus, E., Kluding, H., 1984. Mutations affecting the pattern of the larval cuticle in Drosophila melanogaster. I. Zygotic loci on the second chromosome. Wilhelm Roux Arch. Dev. Biol. 193, 267–282.

Nystrom, S.L., Niederhuber, M.J., McKay, D.J., 2020. Expression of E93 provides an instructive cue to control dynamic enhancer activity and chromatin accessibility during development. Development 147, dev181909.

Ogihara, M.H., Hikiba, J., Iga, M., Kataoka, H., 2015. Negative regulation of juvenile hormone analog for ecdysteroidogenic enzymes. J. Insect Physiol. 80, 42–47.

Ohhara, Y., Shimada-Niwa, Y., Niwa, R., Kayashima, Y., Hayashi, Y., Akagi, K., Ueda, H., Yamakawa-Kobayashi, K., Kobayashi, S., 2015. Autocrine regulation of ecdysone synthesis by beta3-octopamine receptor in the prothoracic gland is essential for Drosophila metamorphosis. Proc. Natl. Acad. Sci. U. S. A. 112, 1452–1457.

Ohhara, Y., Kobayashi, S., Yamanaka, N., 2017. Nutrient-dependent endocycling in steroidogenic tissue dictates timing of metamorphosis in Drosophila melanogaster. PLoS Genet. 13, e1006583.

Okamoto, N., Yamanaka, N., 2020. Steroid hormone entry into the brain requires a membrane transporter in Drosophila. Curr. Biol. 30, 359–366.e3.

Okamoto, N., Yamanaka, N., Satake, H., Saegusa, H., Kataoka, H., Mizoguchi, A., 2009a. An ecdysteroid-inducible insulin-like growth factor-like peptide regulates adult development of the silkmoth Bombyx mori. FEBS J. 276, 1221–1232.

Okamoto, N., Yamanaka, N., Yagi, Y., Nishida, Y., Kataoka, H., O'Connor, M.B., Mizoguchi, A., 2009b. A fat body-derived IGF-like peptide regulates postfeeding growth in Drosophila. Dev. Cell 17, 885–891.

Okamoto, N., Viswanatha, R., Bittar, R., Li, Z., Haga-Yamanaka, S., Perrimon, N., Yamanaka, N., 2018. A membrane transporter is required for steroid hormone uptake in drosophila. Dev. Cell 47, 294–305.e7.

Ono, H., Rewitz, K.F., Shinoda, T., Itoyama, K., Petryk, A., Rybczynski, R., Jarcho, M., Warren, J.T., Marqués, G., Shimell, M.J., Gilbert, L.I., O'Connor, M.B., 2006. Spook and Spookier code for stage-specific components of the ecdysone biosynthetic pathway in Diptera. Dev. Biol. 298, 555–570.

Ou, Q., Magico, A., King-Jones, K., 2011. Nuclear receptor DHR4 controls the timing of steroid hormone pulses during Drosophila development. PLoS Biol. 9, e1001160.

Ou, Q., Zeng, J., Yamanaka, N., Brakken-Thal, C., O'Connor, M.B., King-Jones, K., 2016. The insect prothoracic gland as a model for steroid hormone biosynthesis and regulation. Cell Rep. 16, 247–262.

Pan, X., Neufeld, T.P., O'Connor, M.B., 2019. A tissue- and temporal-specific autophagic switch controls drosophila pre-metamorphic nutritional checkpoints. Curr. Biol. 29, 2840–2851.e4.

Pan, X., Connacher, R.P., O'Connor, M.B., 2020. Control of the insect metamorphic transition by ecdysteroid production and secretion. Curr. Opin. Insect. Sci. 43, 11–20.

Parthasarathy, R., Sheng, Z., Sun, Z., Palli, S.R., 2010. Ecdysteroid regulation of ovarian growth and oocyte maturation in the red flour beetle, Tribolium castaneum. Insect Biochem. Mol. Biol. 40, 429–439.

Pener, M.P., Dhadialla, T.S., 2012. An overview of insect growth disruptors: applied aspects. In: Dhadialla, T.S. (Ed.), Advances in Insect Physiology. vol. 43. Academic Press, London, pp. 1–162.

Petryk, A., Warren, J.T., Marqués, G., Jarcho, M.P., Gilbert, L.I., Kahler, J., Parvy, J.P., Li, Y., Dauphin-Villemant, C., O'Connor, M.B., 2003. Shade is the Drosophila P450 enzyme that mediates the hydroxylation of ecdysone to the steroid insect molting hormone 20-hydroxyecdysone. Proc. Natl. Acad. Sci. U. S. A. 100, 13773–13778.

Pondeville, E., Maria, A., Jacques, J.C., Bourgouin, C., Dauphin-Villemant, C., 2008. Anopheles gambiae males produce and transfer the vitellogenic steroid hormone 20-hydroxyecdysone to females during mating. Proc. Natl. Acad. Sci. U. S. A. 105, 19631–19636.

Raikhel, A.S., Brown, M.R., Belles, X., 2005. Hormonal control of reproductive processes. In: Gilbert, L.I., Iatrou, S.S., Gill, S.S. (Eds.), Comprehensive Molecular Insect Science. vol. 3. Elsevier, Oxford, pp. 433–492.

Redfern, C.P., 1984. Evidence for the presence of makisterone A in Drosophila larvae and the secretion of 20-deoxymakisterone A by the ring gland. Proc. Natl. Acad. Sci. U. S. A. 81, 5643–5647.

Regan, J.C., Brandao, A.S., Leitao, A.B., Mantas Dias, A.R., Sucena, E., Jacinto, A., Zaidman-Remy, A., 2013. Steroid hormone signaling is essential to regulate innate immune cells and fight bacterial infection in Drosophila. PLoS Pathog. 9, e1003720.

Rewitz, K.F., Rybczynski, R., Warren, J.T., Gilbert, L.I., 2006. The Halloween genes code for cytochrome P450 enzymes mediating synthesis of the insect moulting hormone. Biochem. Soc. Trans. 34, 1256–1260.
Rewitz, K.F., Yamanaka, N., Gilbert, L.I., O'Connor, M.B., 2009. The insect neuropeptide PTTH activates receptor tyrosine kinase Torso to initiate metamorphosis. Science 326, 1403–1405.
Rewitz, K.F., Yamanaka, N., O'Connor, M.B., 2010. Steroid hormone inactivation is required during the juvenile-adult transition in Drosophila. Dev. Cell 19, 895–902.
Reynolds, R.A., Kwon, H., Smith, R.C., 2020. 20-hydroxyecdysone primes innate immune responses that limit bacterial and malarial parasite survival in anopheles gambiae. mSphere 5, e00983-19.
Richard, D.S., Gilbert, L.I., 1991. Reversible juvenile hormone inhibition of ecdysteroid and juvenile hormone synthesis by the ring gland of Drosophila melanogaster. Experientia 47, 1063–1066.
Richard, D.S., Watkins, N.L., Serafin, R.B., Gilbert, L.I., 1998. Ecdysteroids regulate yolk protein uptake by Drosophila melanogaster. J. Insect Physiol. 44, 637–644.
Riddiford, L.M., 1985. Hormone action at the cellular level. In: Kerkut, G.A., Gilbert, L.I. (Eds.), Comprehensive Insect Physiology, Biochemistry and Pharmacology. Pergamon Press, Oxford, pp. 37–84.
Riddiford, L.M., 2012. How does juvenile hormone control insect metamorphosis and reproduction? Gen. Comp. Endocrinol. 179, 477–484.
Riddiford, L.M., Cherbas, P., Truman, J.W., 2000. Ecdysone receptors and their biological actions. Vitam. Horm. 60, 1–73.
Robinow, S., Talbot, W.S., Hogness, D.S., Truman, J.W., 1993. Programmed cell death in the Drosophila CNS is ecdysone-regulated and coupled with a specific ecdysone receptor isoform. Development 119, 1251–1259.
Robinow, S., Draizen, T.A., Truman, J.W., 1997. Genes that induce apoptosis: transcriptional regulation in identified, doomed neurons of the Drosophila CNS. Dev. Biol. 190, 206–213.
Robinson, P.D., Morgan, E.D., Wilson, I.D., Lafont, R., 1987. The metabolism of ingested and injected [3H]ecdysone by final instar larvae of Heliothis armigera. Physiol. Entomol. 12, 321–330.
Rodenfels, J., Lavrynenko, O., Ayciriex, S., Sampaio, J.L., Carvalho, M., Shevchenko, A., Eaton, S., 2014. Production of systemically circulating Hedgehog by the intestine couples nutrition to growth and development. Genes Dev. 28, 2636–2651.
Romani, P., Bernardi, F., Hackney, J., Dobens, L., Gargiulo, G., Cavaliere, V., 2009. Cell survival and polarity of Drosophila follicle cells require the activity of ecdysone receptor B1 isoform. Genetics 181, 165–175.
Roy, S., Saha, T.T., Zou, Z., Raikhel, A.S., 2018. Regulatory pathways controlling female insect reproduction. Annu. Rev. Entomol. 63, 489–511.
Rus, F., Flatt, T., Tong, M., Aggarwal, K., Okuda, K., Kleino, A., Yates, E., Tatar, M., Silverman, N., 2013. Ecdysone triggered PGRP-LC expression controls Drosophila innate immunity. EMBO J. 32, 1626–1638.
Sakurai, S., Okuda, M., Ohtaki, T., 1989a. Juvenile hormone inhibits ecdysone secretion and responsiveness to prothoracicotropic hormone in prothoracic glands of Bombyx mori. Gen. Comp. Endocrinol. 75, 222–230.
Sakurai, S., Warren, J.T., Gilbert, L.I., 1989b. Mediation of ecdysone synthesis in Manduca sexta by a hemolymph enzyme. Arch. Insect Biochem. Physiol. 10, 179–197.
Schubiger, M., Wade, A.A., Carney, G.E., Truman, J.W., Bender, M., 1998. Drosophila EcR-B ecdysone receptor isoforms are required for larval molting and for neuron remodeling during metamorphosis. Development 125, 2053–2062.

Schwartz, L.M., Truman, J.W., 1982. Peptide and steroid regulation of muscle degeneration in an insect. Science 215, 1420–1421.
Scott, R.L., Diao, F., Silva, V., Park, S., Luan, H., Ewer, J., White, B.H., 2020. Non-canonical eclosion hormone-expressing cells regulate Drosophila Ecdysis. iScience 23, 101108.
Selcho, M., Millan, C., Palacios-Munoz, A., Ruf, F., Ubillo, L., Chen, J., Bergmann, G., Ito, C., Silva, V., Wegener, C., Ewer, J., 2017. Central and peripheral clocks are coupled by a neuropeptide pathway in Drosophila. Nat. Commun. 8, 15563.
Shafiei, M., Moczek, A.P., Nijhout, H.F., 2001. Food availability controls the onset of metamorphosis in the dung beetle Onthophagus taurus (Coleoptera: Scarabaeidae). Physiol. Entomol. 26, 173–180.
Sharma, V., Pandey, A.K., Kumar, A., Misra, S., Gupta, H.P.K., Gupta, S., Singh, A., Buehner, N.A., Ram, K.R., 2017. Functional male accessory glands and fertility in Drosophila require novel ecdysone receptor. PLoS Genet. 13, e1006893.
Shimada-Niwa, Y., Niwa, R., 2014. Serotonergic neurons respond to nutrients and regulate the timing of steroid hormone biosynthesis in Drosophila. Nat. Commun. 5, 5778.
Shimell, M., Pan, X., Martin, F.A., Ghosh, A.C., Leopold, P., O'Connor, M.B., Romero, N.M., 2018. Prothoracicotropic hormone modulates environmental adaptive plasticity through the control of developmental timing. Development 145, dev159699.
Shinbo, H., Happ, G.M., 1989. Effects of ecdysteroids on the growth of the post-testicular reproductive organs in the silkworm, Bombyx mori. J. Insect Physiol. 35, 855–864.
Sieber, M.H., Spradling, A.C., 2015. Steroid signaling establishes a female metabolic state and regulates SREBP to control oocyte lipid accumulation. Curr. Biol. 25, 993–1004.
Simpson, S.J., Sword, G.A., Lo, N., 2011. Polyphenism in insects. Curr. Biol. 21, R738–R749.
Slaidina, M., Delanoue, R., Gronke, S., Partrigde, L., Léopold, P., 2009. A Drosophila insulin-like peptide promotes growth during nonfeeding states. Dev. Cell 17, 874–884.
Smagghe, G., Gomez, L.E., Dhadialla, T.S., 2012. Bisacylhydrazine insecticides for selective pest control. In: Dhadialla, T.S. (Ed.), Advances in Insect Physiology. vol. 43. Academic Press, London, pp. 163–249.
Smith, W.A., Lamattina, A., Collins, M., 2014. Insulin signaling pathways in lepidopteran ecdysone secretion. Front. Physiol. 5, 19.
Song, Q., Sun, X., Jin, X.Y., 2003. 20E-regulated USP expression and phosphorylation in Drosophila melanogaster. Insect Biochem. Mol. Biol. 33, 1211–1218.
Sonobe, H., Yamada, R., 2004. Ecdysteroids during early embryonic development in silkworm Bombyx mori: metabolism and functions. Zoolog. Sci. 21, 503–516.
Soriano, I., Riley, I.T., Potter, M.J., Bowers, W.S., 2004. Phytoecdysteroids: a novel defense against plant-parasitic nematodes. J. Chem. Ecol. 30, 1885–1899.
Spindler, K.D., Hönl, C., Tremmel, C., Braun, S., Ruff, H., Spindler-Barth, M., 2009. Ecdysteroid hormone action. Cell. Mol. Life Sci. 66, 3837–3850.
Srivastava, D.P., Yu, E.J., Kennedy, K., Chatwin, H., Reale, V., Hamon, M., Smith, T., Evans, P.D., 2005. Rapid, nongenomic responses to ecdysteroids and catecholamines mediated by a novel Drosophila G-protein-coupled receptor. J. Neurosci. 25, 6145–6155.
Steel, C.G.H., Bollenbacher, W.E., Smith, S.L., Gilbert, L.I., 1982. Haemolymph ecdysteroid titres during larval–adult development in Rhodnius prolixus: correlations with molting hormone action and brain neurosecretory cell activity. J. Insect Physiol. 28, 519–525.
Storelli, G., Defaye, A., Erkosar, B., Hols, P., Royet, J., Leulier, F., 2011. Lactobacillus plantarum promotes Drosophila systemic growth by modulating hormonal signals through TOR-dependent nutrient sensing. Cell Metab. 14, 403–414.

Strambi, A., Strambi, C., Röseler, P.F., Röseler, I., 1984. Simultaneous determination of juvenile hormone and ecdysteroid titers in the hemolymph of bumblebee prepupae (Bombus hypnorum and Bombus terrestris). Gen. Comp. Endocrinol. 55, 83–88.
Sun, W., Shen, Y.H., Qi, D.W., Xiang, Z.H., Zhang, Z., 2012. Molecular cloning and characterization of Ecdysone oxidase and 3-dehydroecdysone-3α-reductase involved in the ecdysone inactivation pathway of silkworm, Bombyx mori. Int. J. Biol. Sci. 8, 125–138.
Sun, W., Shen, Y.H., Zhou, L.X., Zhang, Z., 2016. Ecdysone titer determined by 3DE-3beta-reductase enhances the immune response in the silkworm. J. Immunol. 196, 1646–1654.
Sun, W., Wang, C.F., Zhang, Z., 2017. Transcription factor E74A affects the ecdysone titer by regulating the expression of the EO gene in the silkworm, Bomby mori. Biochim. Biophys. Acta 1861, 551–558.
Suzzoni, J.P., Passera, L., Strambi, A., 1980. Ecdysteroid titre and caste determination in the ant, Pheidole pallidula (Nyl.) (Hymenoptera: Formicidae). Experientia 36, 1228–1229.
Suzzoni, J.P., Passera, L., Strambi, A., 1983. Ecdysteroid production during caste differentiation in larvae of the ant, Plagiolepis pygmaea. Physiol. Entomol. 8, 93–96.
Swevers, L., 2019. An update on ecdysone signaling during insect oogenesis. Curr. Opin. Insect. Sci. 31, 8–13.
Swevers, L., Iatrou, K., 2003. The ecdysone regulatory cascade and ovarian development in lepidopteran insects: insights from the silkmoth paradigm. Insect Biochem. Mol. Biol. 33, 1285–1297.
Syed, M.H., Mark, B.J., Doe, C.Q., 2017a. Steroid hormone induction of temporal gene expression in Drosophila brain neuroblasts generates neuronal and glial diversity. Elife 6, e26287.
Syed, M.H., Mark, B.J., Doe, C.Q., 2017b. Playing well with others: extrinsic cues regulate neural progenitor temporal identity to generate neuronal diversity. Trends Genet. 33, 933–942.
Takeuchi, H., Chen, J., O'Reilly, D.R., Rees, H.H., Turner, P.C., 2000. Regulation of ecdysteroid signalling: molecular cloning, characterization and expression of 3-dehydroecdysone 3α-reductase, a novel eukaryotic member of the short-chain dehydrogenases/reductases superfamily from the cotton leafworm, Spodoptera littoralis. Biochem. J. 349, 239–245.
Takeuchi, H., Chen, J., O'Reilly, D.R., Turner, P.C., Rees, H.H., 2001. Regulation of ecdysteroid signaling: cloning and characterization of ecdysone oxidase: a novel steroid oxidase from the cotton leafworm, Spodoptera littoralis. J. Biol. Chem. 276, 26819–26828.
Talamillo, A., Herboso, L., Pirone, L., Pérez, C., González, M., Sánchez, J., Mayor, U., Lopitz-Otsoa, F., Rodriguez, M.S., Sutherland, J.D., Barrio, R., 2013. Scavenger receptors mediate the role of SUMO and Ftz-f1 in Drosophila steroidogenesis. PLoS Genet. 9, e1003473.
Terashima, J., Bownes, M., 2004. Translating available food into the number of eggs laid by Drosophila melanogaster. Genetics 167, 1711–1719.
Terashima, J., Takaki, K., Sakurai, S., Bownes, M., 2005. Nutritional status affects 20-hydroxyecdysone concentration and progression of oogenesis in Drosophila melanogaster. J. Endocrinol. 187, 69–79.
Texada, M.J., Malita, A., Christensen, C.F., Dall, K.B., Faergeman, N.J., Nagy, S., Halberg, K.A., Rewitz, K., 2019a. Autophagy-mediated cholesterol trafficking controls steroid production. Dev. Cell 48, 659–671.e4.
Texada, M.J., Malita, A., Rewitz, K., 2019b. Autophagy regulates steroid production by mediating cholesterol trafficking in endocrine cells. Autophagy 15, 1478–1480.
Texada, M.J., Koyama, T., Rewitz, K., 2020. Regulation of body size and growth control. Genetics 216, 269–313.

Thomas, H.E., Stunnenberg, H.G., Stewart, A.F., 1993. Heterodimerization of the Drosophila ecdysone receptor with retinoid X receptor and ultraspiracle. Nature 362, 471–475.

Thompson, M.J., Svoboda, J.A., Lozano, R., Wilzer, K.R., 1988. Profile of free and conjugated ecdysteroids and ecdysteroid acids during embryonic development of Manduca sexta (L.) following maternal incorporation of [14C]cholesterol. Arch. Insect Biochem. Physiol. 7, 157–172.

Thummel, C.S., 1996. Flies on steroids—Drosophila metamorphosis and the mechanisms of steroid hormone action. Trends Genet. 12, 306–310.

Tillman, J.A., Seybold, S.J., Jurenka, R.A., Blomquist, G.J., 1999. Insect pheromones—an overview of biosynthesis and endocrine regulation. Insect Biochem. Mol. Biol. 29, 481–514.

Tissot, M., Stocker, R.F., 2000. Metamorphosis in Drosophila and other insects: the fate of neurons throughout the stages. Prog. Neurobiol. 62, 89–111.

Truman, J.W., 1996. Steroid receptors and nervous systems metamorphosis in insects. Dev. Neurosci. 18, 87–101.

Truman, J.W., 2005. Hormonal control of insect ecdysis: endocrine cascades for coordinating behavior with physiology. Vitam. Horm. 73, 1–30.

Truman, J.W., 2019. The evolution of insect metamorphosis. Curr. Biol. 29, R1252–R1268.

Truman, J.W., Reiss, S.E., 1976. Dendritic reorganization of an identified motoneuron during metamorphosis of the tobacco hornworm moth. Science 15, 477–479.

Truman, J.W., Riddiford, L.M., 2019. The evolution of insect metamorphosis: a developmental and endocrine view. Philos. Trans. R. Soc. Lond. B Biol. Sci. 374, 20190070.

Truman, J.W., Schwartz, L.M., 1984. Steroid regulation of neuronal death in the moth nervous system. J. Neurosci. 4, 274–280.

Tsai, C.C., Kao, H.Y., Yao, T.P., McKeown, M., Evans, R.M., 1999. SMRTER, a Drosophila nuclear receptor coregulator, reveals that EcR-mediated repression is critical for development. Mol. Cell 4, 175–186.

Uchibori-Asano, M., Kayukawa, T., Sezutsu, H., Shinoda, T., Daimon, T., 2017. Severe developmental timing defects in the prothoracicotropic hormone (PTTH)-deficient silkworm, Bombyx mori. Insect Biochem. Mol. Biol. 87, 14–25.

Ureña, E., Manjón, C., Franch-Marro, X., Martín, D., 2014. Transcription factor E93 specifies adult metamorphosis in hemimetabolous and holometabolous insects. Proc. Natl. Acad. Sci. U. S. A. 111, 7024–7029.

Ureña, E., Chafino, S., Manjón, C., Franch-Marro, X., Martín, D., 2016. The occurrence of the holometabolous pupal stage requires the interaction between E93, Krüppel homolog 1 and Broad-Complex. PLoS Genet. 12, e1006020.

Uryu, O., Ameku, T., Niwa, R., 2015. Recent progress in understanding the role of ecdysteroids in adult insects: germline development and circadian clock in the fruit fly Drosophila melanogaster. Zool. Lett. 1, 32.

Uyehara, C.M., McKay, D.J., 2019. Direct and widespread role for the nuclear receptor EcR in mediating the response to ecdysone in Drosophila. Proc. Natl. Acad. Sci. U. S. A. 116, 9893–9902.

Uyehara, C.M., Nystrom, S.L., Niederhuber, M.J., Leatham-Jensen, M., Ma, Y., Buttitta, L.A., McKay, D.J., 2017. Hormone-dependent control of developmental timing through regulation of chromatin accessibility. Genes Dev. 31, 862–875.

Vafopoulou, X., Steel, C.G., 1991. Circadian regulation of synthesis of ecdysteroids by prothoracic glands of the insect Rhodnius prolixus: evidence of a dual oscillator system. Gen. Comp. Endocrinol. 83, 27–34.

Vafopoulou, X., Steel, C.G., Terry, K.L., 2007. Neuroanatomical relations of prothoracicotropic hormone neurons with the circadian timekeeping system in the brain of larval and adult Rhodnius prolixus (Hemiptera). J. Comp. Neurol. 503, 511–524.
Vallejo, D.M., Juarez-Carreño, S., Bolivar, J., Morante, J., Dominguez, M., 2015. A brain circuit that synchronizes growth and maturation revealed through Dilp8 binding to Lgr3. Science 350, aac6767.
Verma, P., Cohen, S.M., 2015. miR-965 controls cell proliferation and migration during tissue morphogenesis in the Drosophila abdomen. Elife 4, e07389.
Veverytsa, L., Allan, D.W., 2012. Temporally tuned neuronal differentiation supports the functional remodeling of a neuronal network in Drosophila. Proc. Natl. Acad. Sci. U. S. A. 109, E748–E756.
Vogel, K.J., Brown, M.R., Strand, M.R., 2015. Ovary ecdysteroidogenic hormone requires a receptor tyrosine kinase to activate egg formation in the mosquito Aedes aegypti. Proc. Natl. Acad. Sci. U. S. A. 112, 5057–5062.
Walkiewicz, M.A., Stern, M., 2009. Increased insulin/insulin growth factor signaling advances the onset of metamorphosis in Drosophila. PLoS One 4, e5072.
Wang, S., Wang, J., Sun, Y., Song, Q., Li, S., 2012. PKC-mediated USP phosphorylation at Ser35 modulates 20-hydroxyecdysone signaling in Drosophila. J. Proteome Res. 11, 6187–6196.
Wang, D., Zhao, W.L., Cai, M.J., Wang, J.X., Zhao, X.F., 2015. G-protein-coupled receptor controls steroid hormone signaling in cell membrane. Sci. Rep. 5, 8675.
Wang, Z., Lee, G., Vuong, R., Park, J.H., 2019. Two-factor specification of apoptosis: TGF-beta signaling acts cooperatively with ecdysone signaling to induce cell- and stage-specific apoptosis of larval neurons during metamorphosis in Drosophila melanogaster. Apoptosis 24, 972–989.
Warren, J.T., Petryk, A., Marqués, G., Jarcho, M., Parvy, J.P., Dauphin-Villemant, C., O'Connor, M.B., Gilbert, L.I., 2002. Molecular and biochemical characterization of two P450 enzymes in the ecdysteroidogenic pathway of Drosophila melanogaster. Proc. Natl. Acad. Sci. U. S. A. 99, 11043–11048.
Warren, J.T., Petryk, A., Marqués, G., Parvy, J.P., Shinoda, T., Itoyama, K., Kobayashi, J., Jarcho, M., Li, Y., O'Connor, M.B., Dauphin-Villemant, C., Gilbert, L.I., 2004. Phantom encodes the 25-hydroxylase of Drosophila melanogaster and Bombyx mori: a P450 enzyme critical in ecdysone biosynthesis. Insect Biochem. Mol. Biol. 34, 991–1010.
Weeks, J.C., Truman, J.W., 1985. Independent steroid control of the fates of metamorphosis and their muscles during insect metamorphosis. J. Neurosci. 5, 2290–2300.
Whitehead, L., 1989. Ecdysteroid carrier proteins. In: Koolman, J. (Ed.), Ecdysone: From Chemistry to Mode of Action. Thieme Medical Publishers, Stuttgart, pp. 232–244.
Wicker, C., Jallon, J.M., 1995. Hormonal control of sex pheromone biosynthesis in Drosophila melanogaster. J. Insect Physiol. 41, 65–70.
Wieschaus, E., Nüsslein-Volhard, C., 2016. The Heidelberg screen for pattern mutants of Drosophila: a personal account. Annu. Rev. Cell Dev. Biol. 32, 1–46.
Wieschaus, E., Nüsslein-Volhard, C., Jürgens, G., 1984. Mutations affecting the pattern of the larval cuticle in Drosophila melanogaster. III. Zygotic loci on the X-chromosome and the fourth chromosome. Wilhelm Roux Arch. Dev. Biol. 193, 296–307.
Wigglesworth, V.B., 1934. Factors controlling moulting and 'metamorphosis' in an insect. Nature 133, 725–726.
Williams, C.M., 1947. Physiology of insect diapause. II. Interaction between the pupal brain and prothoracic glands in the metamorphosis of the giant silkworm, Platysamia cecropia. Biol. Bull. 93, 89–98.
Williams, D.W., Truman, J.W., 2005. Cellular mechanisms of dendrite pruning in Drosophila: insights from in vivo time-lapse of remodeling dendritic arborizing sensory neurons. Development 132, 3631–3642.

Williams, D.R., Fisher, M.J., Rees, H.H., 2000. Characterization of ecdysteroid 26-hydroxylase: an enzyme involved in molting hormone inactivation. Arch. Biochem. Biophys. 376, 389–398.

Winbush, A., Weeks, J.C., 2011. Steroid-triggered, cell-autonomous death of a Drosophila motoneuron during metamorphosis. Neural Dev. 6, 15.

Wu, Q., Jiang, Z.J., Bai, C., 1994. The role of ecdysteroids in the reproductive diapause of Dermacentor niveus Neumann. Entomol. Sinica 1, 164–171.

Xu, J., Raman, C., Zhu, F., Tan, A., Palli, S.R., 2012. Identification of nuclear receptors involved in regulation of male reproduction in the red flour beetle, Tribolium castaneum. J. Insect Physiol. 58, 710–717.

Xu, L.C., Nunes, C., Wang, V.R., Saito, A., Chen, T., Basak, P., Chang, J.J., Koyama, T., Suzuki, Y., 2020a. Distinct nutritional and endocrine regulation of prothoracic gland activities underlies divergent life history strategies in Manduca sexta and Drosophila melanogaster. Insect Biochem. Mol. Biol. 119, 103335.

Xu, T., Jiang, X., Denton, D., Kumar, S., 2020b. Ecdysone controlled cell and tissue deletion. Cell Death Differ. 27, 1–14.

Yamada, N., Kataoka, H., Mizoguchi, A., 2017. Myosuppressin is involved in the regulation of pupal diapause in the cabbage army moth Mamestra brassicae. Sci. Rep. 7, 41651.

Yamada, T., Hironaka, K.I., Habara, O., Morishita, Y., Nishimura, T., 2020. A developmental checkpoint directs metabolic remodelling as a strategy against starvation in Drosophila. Nat. Metab. 2, 1096–1112.

Yamanaka, N., Hua, Y.J., Mizoguchi, A., Watanabe, K., Niwa, R., Tanaka, Y., Kataoka, H., 2005. Identification of a novel prothoracicostatic hormone and its receptor in the silkworm Bombyx mori. J. Biol. Chem. 280, 14684–14690.

Yamanaka, N., Žitňan, D., Kim, Y.J., Adams, M.E., Hua, Y.J., Suzuki, Y., Suzuki, M., Suzuki, A., Satake, H., Mizoguchi, A., Asaoka, K., Tanaka, Y., Kataoka, H., 2006. Regulation of insect steroid hormone biosynthesis by innervating peptidergic neurons. Proc. Natl. Acad. Sci. U. S. A. 103, 8622–8627.

Yamanaka, N., Honda, N., Osato, N., Niwa, R., Mizoguchi, A., Kataoka, H., 2007. Differential regulation of ecdysteroidogenic P450 gene expression in the silkworm, Bombyx mori. Biosci. Biotechnol. Biochem. 71, 2808–2814.

Yamanaka, N., Hua, Y.J., Roller, L., Spalovská-Valachová, I., Mizoguchi, A., Kataoka, H., Tanaka, Y., 2010. Bombyx prothoracicostatic peptides activate the sex peptide receptor to regulate ecdysteroid biosynthesis. Proc. Natl. Acad. Sci. U. S. A. 107, 2060–2065.

Yamanaka, N., Roller, L., Zitňan, D., Satake, H., Mizoguchi, A., Kataoka, H., Tanaka, Y., 2011. Bombyx orcokinins are brain-gut peptides involved in the neuronal regulation of ecdysteroidogenesis. J. Comp. Neurol. 519, 238–246.

Yamanaka, N., Rewitz, K.F., O'Connor, M.B., 2013. Ecdysone control of developmental transitions: lessons from Drosophila research. Annu. Rev. Entomol. 58, 497–516.

Yamanaka, N., Marqués, G., O'Connor, M.B., 2015. Vesicle-mediated steroid hormone secretion in Drosophila melanogaster. Cell 163, 907–919.

Yang, H.J., Wang, M.X., Zhang, P., Sabhat, A., Malik, F.A., Bhaskar, R., Zhou, F., Li, X.H., Hu, J.B., Sun, C.G., Niu, Y.S., Miao, Y.G., 2011. Cloning and characterization of the Bombyx mori ecdysone oxidase. Arch. Insect Biochem. Physiol. 78, 17–29.

Yaniv, S.P., Schuldiner, O., 2016. A fly's view of neuronal remodeling. Wiley Interdiscip. Rev. Dev. Biol. 5, 618–635.

Yao, T.P., Segraves, W.A., Oro, A.E., McKeown, M., Evans, R.M., 1992. Drosophila ultraspiracle modulates ecdysone receptor function via heterodimer formation. Cell 71, 63–72.

Yao, T.P., Forman, B.M., Jiang, Z., Cherbas, L., Chen, J.D., McKeown, M., Cherbas, P., Evans, R.M., 1993. Functional ecdysone receptor is the product of EcR and Ultraspiracle genes. Nature 366, 476–479.

Yoshinari, Y., Kurogi, Y., Ameku, T., Niwa, R., 2019. Endocrine regulation of female germline stem cells in the fruit fly Drosophila melanogaster. Curr. Opin. Insect. Sci. 31, 14–19.

Yoshiyama, T., Namiki, T., Mita, K., Kataoka, H., Niwa, R., 2006. Neverland is an evolutionally conserved Rieske-domain protein that is essential for ecdysone synthesis and insect growth. Development 133, 2565–2574.

Yoshiyama-Yanagawa, T., Enya, S., Shimada-Niwa, Y., Yaguchi, S., Haramoto, Y., Matsuya, T., Shiomi, K., Sasakura, Y., Takahashi, S., Asashima, M., Kataoka, H., Niwa, R., 2011. The conserved Rieske oxygenase DAF-36/Neverland is a novel cholesterol-metabolizing enzyme. J. Biol. Chem. 286, 25756–25762.

Yu, F., Schuldiner, O., 2014. Axon and dendrite pruning in Drosophila. Curr. Opin. Neurobiol. 27, 192–198.

Zeng, J., Huynh, N., Phelps, B., King-Jones, K., 2020. Snail synchronizes endocycling in a TOR-dependent manner to coordinate entry and escape from endoreplication pausing during the Drosophila critical weight checkpoint. PLoS Biol. 18, e3000609.

Zhang, Z., Palli, S.R., 2009. Identification of a cis-regulatory element required for 20-hydroxyecdysone enhancement of antimicrobial peptide gene expression in Drosophila melanogaster. Insect Mol. Biol. 18, 595–605.

Zhang, C., Robinson, B.S., Xu, W., Yang, L., Yao, B., Zhao, H., Byun, P.K., Jin, P., Veraksa, A., Moberg, K.H., 2015. The ecdysone receptor coactivator Taiman links Yorkie to transcriptional control of germline stem cell factors in somatic tissue. Dev. Cell 34, 168–180.

Zhang, B., Sato, K., Yamamoto, D., 2018a. Ecdysone signaling regulates specification of neurons with a male-specific neurite in Drosophila. Biol. Open. 7, 1–8.

Zhang, T., Song, W., Li, Z., Qian, W., Wei, L., Yang, Y., Wang, W., Zhou, X., Meng, M., Peng, J., Xia, Q., Perrimon, N., Cheng, D., 2018b. Krüppel homolog 1 represses insect ecdysone biosynthesis by directly inhibiting the transcription of steroidogenic enzymes. Proc. Natl. Acad. Sci. U. S. A. 115, 3960–3965.

Zhao, X.F., 2020. G protein-coupled receptors function as cell membrane receptors for the steroid hormone 20-hydroxyecdysone. Cell Commun. Signal 18, 146.

Zhu, X.X., Gfeller, H., Lanzrein, B., 1983. Ecdysteroids during oogenesis in the ovoviviparous cockroach Nauphoeta cinerea. J. Insect Physiol. 29, 225–235.

Zipper, L., Jassmann, D., Burgmer, S., Görlich, B., Reiff, T., 2020. Ecdysone steroid hormone remote controls intestinal stem cell fate decisions via the PPARγ-homolog Eip75B in Drosophila. Elife 9, e55795.

Zirin, J., Cheng, D., Dhanyasi, N., Cho, J., Dura, J.M., VijayRaghavan, K., Perrimon, N., 2013. Ecdysone signaling at metamorphosis triggers apoptosis of Drosophila abdominal muscles. Dev. Biol. 383, 275–284.

Žitňan, D., Adams, M.E., 2012. Neuroendocrine regulation of ecdysis. In: Gilbert, L.I. (Ed.), Insect Endocrinology. Academic Press, London, pp. 253–309.

Žitňan, D., Kingan, T.G., Hermesman, J., Adams, M.E., 1996. Identification of ecdysis-triggering hormone from an epitracheal endocrine system. Science 271, 88–91.

Žitňan, D., Ross, L.S., Žitňanová, I., Hermesman, J.L., Gill, S.S., Adams, M.E., 1999. Steroid induction of a peptide hormone gene leads to orchestration of a defined behavioral sequence. Neuron 23, 523–535.

CHAPTER TWO

A decade with the juvenile hormone receptor

Marek Jindra[a,*], Sarka Tumova[a], Matej Milacek[b], and Lenka Bittova[a]
[a]Biology Center of the Czech Academy of Sciences, Institute of Entomology, Ceske Budejovice, Czech Republic
[b]Department of Molecular Biology and Genetics, University of South Bohemia, Ceske Budejovice, Czech Republic
*Corresponding author: e-mail address: jindra@entu.cas.cz

Contents

Abstract

Juvenile hormone (JH) and ecdysteroids represent equally important nonpeptide signals governing insect reproduction and development. For a considerable time, understanding of JH action lagged behind ecdysteroid research. Arriving with a 20-year delay, the intracellular receptor for JH was finally identified as a ligand-activated bHLH-PAS transcription factor, originally named Methoprene-tolerant (Met), by virtue of the resistance *Drosophila* mutants exhibited to morphogenetic and lethal actions of JH and its mimic methoprene. Systemic RNAi in suitable insect models revealed the anticipated function of Met in preventing metamorphosis and promoting reproduction, thus providing the missing link to the chief roles of JH in insects. That, along with defining the

Advances in Insect Physiology, Volume 60
ISSN 0065-2806
https://doi.org/10.1016/bs.aiip.2021.03.001

JH-binding pocket and the JH-response DNA elements of Met, established the JH receptor (JHR) 10 years ago. After reviewing the functional attributes of the JHR, this chapter will focus on advances in the genetics, cell biology, and biochemistry of the JHR achieved during this past decade. Although hormone receptor function of bHLH-PAS transcription factors is unprecedented, the well-studied mammalian aryl hydrocarbon receptor (AhR) belonging to the same protein family affords functional parallels with the JHR. We can now begin to understand the mechanisms of JHR interaction with the chaperone Hsp90/83, nucleocytoplasmic transport and post-translational regulation by phosphorylation, dimerization with bHLH-PAS partner proteins, and activation by agonist ligands binding to the PAS-B domain. A section is dedicated to current efforts exploiting the JHR as a basis of chemical high-throughput screening, aimed at discovery of novel compounds for environmentally friendly control of insect pests and disease vectors.

1. Introduction

Arthropods possess lipophilic hormones of two major types: steroids (ecdysteroids, mainly represented by ecdysone and its active metabolite 20-hydroxyecdysone; 20E), and the sesquiterpenoid juvenile hormones (JHs). In insects, ecdysteroids and JHs together coordinate the complex post-embryonic development that involves the periodic moulting of larvae and their metamorphosis to adults. While 20E promotes both larval and metamorphic moults, the presence of JH maintains the juvenile character of a moulting larva and prevents it from metamorphosing prematurely, before attaining a required stage and body size (Bellés, 2020; Jindra, 2019; Jindra et al., 2013; Nijhout et al., 2014; Riddiford, 1994; Truman, 2019; Tsang et al., 2020; Yamanaka et al., 2013). Both 20E and JH are also important for reproduction, particularly for vitellogenesis and ovarian maturation, with the exact role of either hormone depending on the insect species (Hansen et al., 2014; Roy et al., 2018; Santos et al., 2019; Swevers, 2019; Wu et al., 2021; Wyatt and Davey, 1996). Long before their chemical identification (Butenandt and Karlson, 1954; Röller et al., 1967), both hormones had been postulated as bloodborne factors emanating from localized organs based on their effects on moulting and metamorphosis of the kissing bug, *Rhodnius prolixus* (Wigglesworth, 1934, 1936, 1940). Besides their general biological importance, both ecdysteroid and JH signalling became targets of chemical intervention, aiming to disrupt the endocrine system in insect pests and disease vectors (Pener and Dhadialla, 2012). JH mimics such as methoprene (Henrick, 2007; Henrick et al., 1973) became the first insecticides of a new generation termed "insect growth regulators" (IGRs) or, better, insect growth disruptors (Minakuchi and Riddiford, 2006; Parthasarathy et al., 2012; Pener and Dhadialla, 2012; Sláma et al., 1974).

Steroid hormones occur universally in arthropods and other animals. They regulate expression of target genes through intracellular receptors of the nuclear receptor (NR) family (Evans and Mangelsdorf, 2014; King-Jones and Thummel, 2005; Mangelsdorf et al., 1995). An active 20E receptor, initially identified in the fruit fly, *Drosophila melanogaster*, is a heterodimer of the ecdysone receptor (EcR) and ultraspiracle (Usp) NR proteins (Koelle et al., 1991; Thomas et al., 1993; Yao et al., 1993) acting on target genes through ecdysone response DNA elements (Cherbas et al., 1991). Work on *Drosophila* mutants revealed the essential requirement for *EcR* and *usp* genes in 20E-dependent processes including moulting, metamorphosis, and oogenesis (Bender et al., 1997; Carney and Bender, 2000; Hall and Thummel, 1998; Henrich et al., 1994; Oro et al., 1992). Three-dimensional structures of the EcR and Usp proteins from several insect species have been resolved through X-ray crystallography (Billas et al., 2003; Carmichael et al., 2005; Iwema et al., 2007; Ren et al., 2014), providing deep mechanistic insights into the hormone-receptor interaction (reviewed by Hill et al., 2012, 2013).

Despite equal importance of 20E and JH signalling in insect biology, the JH field lagged considerably behind the rapidly progressing ecdysteroid research. The molecular mode of JH action had for long remained a conundrum, particularly as the receptor mediating the anti-metamorphic effect of JH could not be identified (Riddiford, 2020). Sesquiterpenoids are structurally unique hormones. Their biosynthesis downstream of the mevalonate pathway involves several steps that are considered specific to arthropods (Goodman and Cusson, 2012; Hiruma and Kaneko, 2013; Qu et al., 2015; Tsang et al., 2020) [although a direct JH precursor with hormonal activity, methyl farnesoate, has been found in annelids (Schenk et al., 2016)]. Unlike steroid hormones with their common gene-regulatory mechanism, no precedent intracellular receptor system was available to guide the quest for JH receptors.

This chapter will briefly recapitulate the milestones en route to identifying JH receptors encoded by the *Methoprene-tolerant* (*Met*) genes and their *Drosophila* ortholog known as *germ cell-expressed* (*gce*) before explaining attributes that define their receptor function. Next, we will summarize what has been learned so far about the functional architecture and molecular interactions of these JH receptor proteins. Finally, we will mention recent attempts to exploit knowledge of the JH receptor

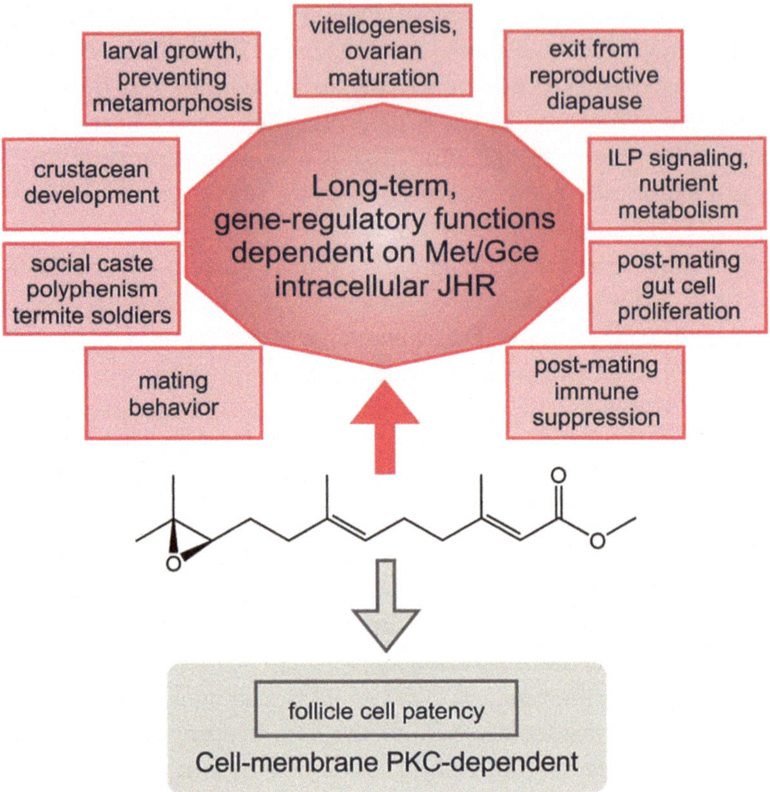

Fig. 1 Examples of processed known to be regulated by JH either through Met/Gce or at the cell-membrane level. A structure of JH III is shown. For citations see Section 3.

molecules towards developing new compounds for manipulating JH signalling and for chemical control of insects.

Throughout the text, Met/Gce will be explicitly or implicitly regarded as an intracellular JH receptor, abbreviated JHR where appropriate. This is to distinguish the gene-regulatory signalling mediated by the intracellular JHR (Fig. 1) from additional effects exerted by JH on as yet unidentified receptor(s) at the cell surface.

2. From JH effects to discovering a JH receptor

Some of the processes regulated by JH are reflected in Fig. 1. The history of research leading to discoveries of JH receptors has been eloquently captured in a recent review by Riddiford (2020), and previously detailed in another chapter of this book series (Dubrovsky and Bernardo, 2014). In this section, we will highlight selected studies, which formed our perception of how an intracellular JHR should function, and those leading to its eventual identification.

2.1 Developmental action of JH: A take-home message from one paper

In an early experiment, Wigglesworth (1958) treated the cuticle of last-instar *Rhodnius* larvae with extracts from male *Hyalophora cecropia* silkmoths (Williams, 1956), a natural source of JH. Prior abrasion of the cuticle ensured that the hormone, applied in oil, would penetrate the cuticle and act on the larval epidermis locally. During the ensuing metamorphic moult, the epidermal cells produced the smooth adult cuticle except in the areas that were previously abraded and exposed to JH, and where the cuticle retained the rough and dark larval character. In this way, even the author's initials (V.B.W.) could be made to appear on the adult, literally printed in the dark larval cuticle type. Similarly, formation of adult structures such as the wings could be blocked by treating the wing pads (Wigglesworth, 1958). In all instances, the hormone had to be delivered early during the last larval instar before the epidermis became committed to metamorphosis.

Viewed from today's perspective of a developmental geneticist, this simple and elegant experiment has had important implications for understanding the mode of JH action. First, while being a systemic, bloodborne signal, JH acts in a cell-autonomous manner. In other words, the signal is not uniformly spread throughout the body. Rather, JH reception is an inherent property of each individual cell. Second, the timely presence of JH alone can block metamorphosis of tissues or body parts. Therefore, JH alters the outcome of a moult induced by ecdysteroids; its presence *vs.* absence, respectively, determines the larval *vs.* adult fate decision in target organs. By cell-autonomously perpetuating the larval developmental programme within an animal undergoing metamorphosis, JH can cause heterochronic phenotypes, wherein treated individuals are spatiotemporal mosaics bearing juvenile and adult structures.

Developmental effects such as those described above are best compatible with a gene-regulatory mechanism, suggesting that JH may alter the genetic programme of recipient cells at the transcriptional level. By analogy to other nonpeptide, lipophilic hormones and morphogens such as steroids or retinoic acid, one would expect JH action to be mediated by an intracellular receptor, likely a ligand-activated transcription factor functionally akin to members of the NR family (King-Jones and Thummel, 2005). The pursuit of such a receptor would eventually occupy the research community for many decades following Wigglesworth's seminal discoveries.

2.2 Mutual benefits of *Drosophila* and complementary insect models

The thriving ecdysteroid field owed a great deal to the well-defined effects of 20E on *Drosophila* genes and development. Transcriptional activity triggered by the hormone could be directly observed *ex vivo* as "puffs" forming on giant salivary gland chromosomes (Ashburner et al., 1974). This facility aided identification and subsequent cloning of 20E-induced genes, including the canonical 20E receptor (EcR) itself (for reviews see Hill et al., 2013; Thummel, 1996). Unrivalled genetic tools have made *Drosophila* the perfect model to study steroid hormone signalling (Yamanaka et al., 2013).

By contrast, the lack of a robust effect of JH on *Drosophila* (Gilbert et al., 2000; Riddiford, 1994) hindered a straightforward approach to uncovering JH signalling effectors and target genes. Unlike in many other insects, ectopic JH treatment in *Drosophila* does not cause extra larval instars and does not prevent metamorphosis of most body parts of the adult fly. However, limited effects on *Drosophila* adult development do occur when exogenous JH is given during a sensitive period shortly after pupariation. These include disrupted abdominal bristle patterns, incomplete rotation of male genitalia, and failure to eclose (Madhavan, 1973). Similarly, ectopic JH application compromises formation of the adult abdominal epidermis. Some JH mimics such as methoprene and pyriproxyfen [now established JH receptor agonists (Jindra and Bittova, 2020)] induce these morphological defects and block adult eclosion more effectively than JH (Riddiford and Ashburner, 1991; Wilson and Fabian, 1986; Zhou and Riddiford, 2002).

The late-stage lethality caused by methoprene was exploited by Wilson and Fabian (1986), who took a forward-genetic approach to finding targets of JH action. Their mutagenesis screen uncovered a single, chromosome X-linked gene, *Methoprene-tolerant* (*Met*), whose loss mitigated the lethal and morphogenetic impact of treatments with methoprene or with native JH III. Curiously, flies hemi- or homozygous for mutant alleles of *Met* were viable and displayed subtle anomalies, namely of the thoracic muscles, nervous system, externally minor eye defects, and reduced oogenesis (Restifo and Wilson, 1998; Wilson and Ashok, 1998; Wilson and Fabian, 1986; Wilson et al., 2006). Later work revealed that the internal visual system in *Met* mutants was disordered due to precocious development of certain photoreceptors (Riddiford et al., 2010) (for a review see Dubrovsky and Bernardo, 2014).

The absence of an obvious, JH-related phenotype discouraged implications that Met might be a JH receptor, even after the gene had been cloned and shown to encode a member of the basic helix-loop-helix (bHLH)/Per-Arnt-Sim (PAS) family of transcription factors (Ashok et al., 1998). There was no precedent of a bHLH-PAS protein acting as a receptor for any known animal hormone. The scepticism did not dramatically change even after an initial indication that the recombinant Met protein-bound JH III with a high affinity (Miura et al., 2005).

To explain the dispensability of *Met*, Wilson and Ashok (1998) speculated that (*A*) JH had no essential role in pre-adult *Drosophila* development, or (*B*) another gene provided a partial functional redundancy, making *Met* nonvital; (*B*) turned out to be correct. *Met* arose *via* a recent duplication from the ancestral gene *gce* (Baumann et al., 2010) in derived flies (two paralogs occur within the Schizophora). As later demonstrated by Abdou et al. (2011), *Met* only becomes vital when *gce* is deleted and *vice versa*. While the null *gce* mutants are also viable, loss of both *Met* and *gce* causes unconditional lethality at the time of pupation (Abdou et al., 2011). Indeed, both *Met* and *gce* encode JH receptors that are functional in live *Drosophila*, and either gene can substitute for the missing paralog when expressed in the *Met gce/Y* double-mutant males (Jindra et al., 2015b). The situation is easier in most other insect species possessing only single orthologs of *Met/gce*, one exception being lepidopterans where the gene has also duplicated (Kayukawa and Shinoda, 2015).

Unlike *Drosophila*, insects such as moths, beetles, or true bugs present "canonical" JH regulation. This means that supernumerary larval instars or extra pupal cuticles can be induced by ectopic treatment with JH agonists and, conversely, precocious metamorphosis ensues when JH signalling is blocked (Jindra et al., 2013). The arrival of RNA interference (RNAi), enabling systemic gene knockdown in whole larvae (Tomoyasu and Denell, 2004), turned the red flour beetle, *Tribolium castaneum*, into a superior model for JH studies. When the single *Tribolium Met* gene was knocked down just prior to pupation, pupae became refractory to the effects of JH agonist treatment and produced normal adults instead of repeating the pupal programme (Konopova and Jindra, 2007, 2008; Minakuchi et al., 2009). Besides their developmental implications, these data also confirmed Wilson's thesis that "methoprene tolerance results from an absence of target gene product" (Wilson and Ashok, 1998).

Fig. 2 Precocious metamorphosis in *T. castaneum*, induced by RNAi-mediated knockdown of *Met* during the fourth larval instar, manifested as pupation after the penultimate (sixth) larval instar and formation of an adult beetle (see also Konopova and Jindra, 2007). The same phenotype results from knocking down *JHAMT* (Minakuchi et al., 2008a). Scale bar, 0.5 mm.

More importantly, however, *Met* RNAi caused *Tribolium* larvae to pupate prematurely, before reaching their final (seventh) instar, and some of the miniature pupae even developed into pygmy adult beetles (Konopova and Jindra, 2007) (Fig. 2). It was this particular experiment, unachievable in *Drosophila*, which supplied the missing evidence that Met indeed executed the natural, developmental function of JH. The phenotype was a perfect copy of JH deficiency itself (Minakuchi et al., 2008a). In their next study on the linden bug, *Pyrrhocoris apterus*, Konopova et al. (2011) extended the anti-metamorphic role of Met to hemimetabolous insects. This finding, followed by work on additional holo- and hemimetabolous species, has indicated that a common molecular endocrine mechanism directs development in all metamorphosing insects (Bellés, 2020; Jindra, 2019; Truman, 2019).

Although the phenotypes produced by removal of Met in *Tribolium* or *Pyrrhocoris* were clearly linked to JH signalling, they did not prove that Met was a JH receptor. Yet these results gave a new impetus to JH research, and the long-sought receptor was soon to be confirmed. As highlighted in the historical review (Riddiford, 2020), the winning combination enabling the progress was that of complementing *Drosophila* genetics with insect models better representing developmental regulation by JH.

3. Functional attributes defining Met/Gce as a JHR

The processes that depend on Met/Gce proteins as effectors of JH signalling are rather long-term and likely based at the genome level (Fig. 1). Before embarking on the discussion of those gene-regulatory functions of JH, let us briefly pause to consider a more instantaneous effect that relies on signalling pathways other than the intracellular JHR. A reduction of volume of ovarian follicle cells, known as patency, enables yolk precursors to pass through the follicle epithelium. Patency had long been known to depend on a Na^+/K^+-ATPase in *Rhodnius* and other insects (Davey, 1996; Sevala and Davey, 1989). A recent study on *Locusta migratoria* showed that JH induced patency by activating the Na^+/K^+ pump *via* protein kinase C (PKC)-dependent phosphorylation (Jing et al., 2018). The phosphorylation was sensitive to a number of chemical inhibitors, further implicating G protein-coupled receptor, receptor tyrosine kinase, inositol trisphosphate receptor, and phospholipase C signalling upstream of PKC. Nevertheless, a hypothetical cell-membrane receptor, through which JH sets in motion the phosphorylation cascade leading to patency, remains to be identified.

3.1 Loss-of-function phenotypes consistent with JH deficiency

3.1.1 Metamorphosis

As alluded to above, a match between phenotypes caused by hormone deficiency and loss of its candidate receptor is prerequisite to establishing an agonist-receptor relationship. For JH and Met, the initial evidence relied on systemic RNAi in *Tribolium*. The precocious metamorphosis seen in *Met* RNAi larvae (Konopova and Jindra, 2007) (Fig. 2) could be phenocopied by knocking down JH acid methyltransferase (JHAMT) (Minakuchi et al., 2008a). JHAMT catalyzes one of the two final steps in JH biosynthesis within the *corpora allata* glands and is necessary to produce an active JH (Shinoda and Itoyama, 2003; Tsang et al., 2020). The other of the two final steps (the order of which depends on insect species) is epoxidation either of farnesoic acid or of the direct JH III precursor, methyl farnesoate (MF) (Helvig et al., 2004; Tsang et al., 2020). Relative to *JHAMT* RNAi, knockdown of the responsible epoxidase CYP15A1 affected

Tribolium metamorphosis only mildly (Minakuchi et al., 2015), possibly because MF exerts a partial JH agonist activity (Bittova et al., 2019; Jindra et al., 2015b; Wen et al., 2015). Precocious adult development consistent with JH deficiency could be induced by RNAi-mediated knockdown of *Met* in hemimetabolans, including the true bugs *P. apterus* and *R. prolixus* (Konopova et al., 2011; Smykal et al., 2014b; Villalobos-Sambucaro et al., 2015), or the cockroach *Blattella germanica* (Lozano and Bellés, 2014).

Phenotypes resulting either from complete loss of *Met* function or from JH deficiency could be rigorously compared when Daimon et al. (2015) generated gene knockout lines in the silkworm, *Bombyx mori*. The silkworm normally undergoes five larval instars before pupating. Null mutants lacking JHAMT (and hence JH) developed through embryogenesis until they entered precocious metamorphosis as third- or fourth-instar larvae. The *JHAMT*$^{-/-}$ larval-pupal intermediates or abnormal pupae produced no adults (Daimon et al., 2015). A similar albeit milder phenotype occurred in the *dimolting* (mod) strain, which lacks the CYP15C1 epoxidase of farnesoic acid. At their third or fourth larval instar, the *Bombyx mod* mutants formed precocious miniature pupae and adults (Daimon et al., 2012). *JHAMT*$^{-/-}$ *mod* double mutants mostly began to metamorphose at the third instar (Daimon et al., 2015).

There are two *Met* paralogs in *Bombyx*, of which *Met1* is vital while *Met2* is dispensable (Daimon et al., 2015). Some of the *Met1*$^{-/-}$ or double-knockout *Met1*$^{-/-}$ *Met2*$^{-/-}$ larvae died during the second moult and none pupated. Importantly, those surviving to the third instar carried patches of pupal cuticle, a hallmark of precocious metamorphosis consistent with JH deficiency. A fusion of maternal genotypes in mosaic progeny led to fourth-instar larvae in which as much as half of the body, consisting of *Met1*$^{-/-}$ cells, displayed the pupal character (Daimon et al., 2015). These genetic data reinforce the notion that Met is a cell-autonomous JH receptor.

The knockout study in *Bombyx* has cemented previous indications that, in contrast to a traditional view, insect larvae of the earliest two instars are incompetent to enter metamorphosis and therefore JH is not required to maintain their juvenile status (Daimon et al., 2012; Feyereisen and Jindra, 2012; Smykal et al., 2014b; Tan et al., 2005). To be precise, at the tissue-autonomous level the absence of JH reception (Met1) is required, but not sufficient, to permit the pupal programme in the first- and second-instar *Bombyx* epidermis (Inui and Daimon, 2017). The dispensability of JH during the earliest larval instars has been further suggested in the *Aedes aegypti* mosquitoes (Zhu et al., 2019). CRISPR/Cas9-induced mutagenesis of *Met* caused lethality of G0 larvae no earlier than before pupation at the final (fourth) instar.

The effects of *Met/gce* loss could be reconciled with JH deficiency also in *Drosophila*. The cell-death gene *grim* proved effective in removing the source of JH when misexpressed in the *corpora allata* cells, embedded within the *Drosophila* larval ring gland (Liu et al., 2009; Riddiford et al., 2010). Depletion of JH resulted in lethality at the time of pupation, thus matching the lethal phase of *Met gce* double mutants (Abdou et al., 2011). As one would expect, supplementing a JH agonist compensated for the genetic allatectomy, but not for loss of JHR genes (Abdou et al., 2011; Riddiford et al., 2010).

3.1.2 Female reproduction

Promoting vitellogenesis and ovarian maturation in insects is another major role for JH (Hansen et al., 2014; Roy et al., 2018; Santos et al., 2019; Wu et al., 2021; Wyatt and Davey, 1996) (Fig. 1). Reduced oviposition was noticeable in *Drosophila Met* or *gce* mutants (Abdou et al., 2011; Wilson and Ashok, 1998). In *Tribolium* adult females, systemic RNAi-mediated knockdown of either *JHAMT* or *Met* decreased expression of a yolk protein precursor vitellogenin (Vg) in the fat body (Sheng et al., 2011). This work has revealed that in the beetles, JH- and Met-dependent induction of the *Vg2* gene involved up-regulated expression of insulin-like peptide (ILP 2 and 3) genes, thus illustrating a synergy between JH and nutrient-dependent insulin signalling during insect female reproduction (for reviews see Hansen et al., 2014; Smykal and Raikhel, 2015; Wu et al., 2021).

Vitellogenesis strictly depends on JH in hemimetabolous insects such as orthopterans, cockroaches, and some hemipterans. Accordingly, the indispensable requirement of Met for vitellogenesis could be demonstrated in *P. apterus*, *R. prolixus*, the bed bug *Cimex lectularius* (all Heteroptera) (Gujar and Palli, 2016; Smykal et al., 2014a; Villalobos-Sambucaro et al., 2015), the locusts *Schistocerca gregaria* and *L. migratoria* (Gijbels et al., 2019; Guo et al., 2014), or the cockroach *Diploptera punctata* (Marchal et al., 2014). In all cases, RNA-mediated knockdown of *Met* partially or completely blocked *Vg* expression and/or ovarian maturation and oogenesis. In *L. migratoria*, whose *corpora allata* are sensitive to chemical inactivation with precocene, the effect of *Met* RNAi could also be matched by means of chemical allatectomy (Guo et al., 2014). Similarly, surgical allatectomy was performed to corroborate the data in adult females of *P. apterus* (Smykal et al., 2014a) (Fig. 3). As in genetically allatectomized *Drosophila*, chemical

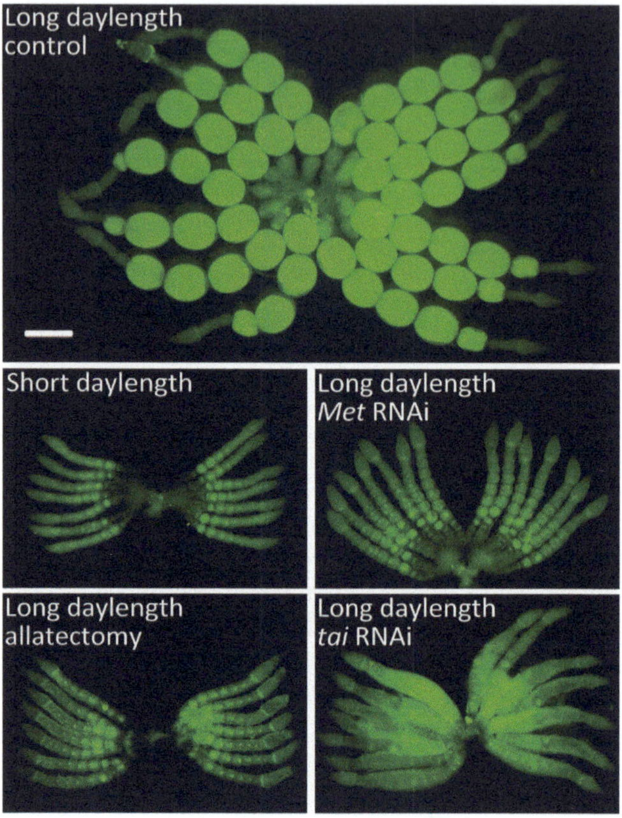

Fig. 3 Oogenesis in the linden bug, *P. apterus*, requires JH and the JHR proteins Met and Tai. Vitellogenic oocytes of reproductive females (top) disappear from the ovarioles during diapause (short daylength) or when the source of JH is removed (allatectomy). RNAi knockdown of *Met* or *tai* blocks oogenesis in formerly reproductive females (Smykal et al., 2014a). The ovaries were stained with Bodipy. Scale bar, 1 mm, applies to all panels. Photographs kindly provided by Vlastimil Smykal.

or surgical depletion of JH could be remedied with JH agonist treatments, but only as long as Met was present.

The *Pyrrhocoris* model offers a condition of natural JH absence during reproductive diapause in adult females (Dolezel, 2015). When their brain perceives shortening daylength, the *corpus allatum* shuts down, causing *Vg* expression in the fat body to cease in preparation for winter. The resulting arrest of oogenesis matches the effects of allatectomy or *Met* knockdown in reproductive females experiencing long daylight (Smykal et al., 2014a) (Fig. 3). Methoprene treatment can override the photoperiodic sensory input and induce *Vg* expression and oogenesis in formerly diapausing, non-reproductive females, provided that *Met* had not been suppressed by prior RNAi treatment (Bajgar et al., 2013; Smykal et al., 2014a). Virtually the same regulation of vitellogenesis applies to a summer diapause in the cabbage beetle, *Colaphellus bowringi*, except in reverse: it is the long daylight associated with adverse summer conditions that informs the reproductive arrest (Liu et al., 2016). The function of Met in JH-mediated exit from a reproductive diapause is therefore conserved across insect orders and eco-physiological strategies.

Extensive evidence for JH- and Met-dependent regulation of genes associated with reproduction has been afforded by studies on *A. aegypti*. In these anautogenous mosquitoes, JH does not directly induce vitellogenesis but rather primes the female for reproduction enabled by the blood meal (Hansen et al., 2014; Roy et al., 2018). Transcriptome analyses of *Aedes* females during post-eclosion, pre-vitellogenic development uncovered large clusters of genes that are differentially regulated in response to the changing JH titre (Saha et al., 2016; Zou et al., 2013). Importantly, expression of many genes positively correlated with high JH titre also required Met, and upstream regions of these genes often contained sequences matching consensus elements capable of binding Met (Zou et al., 2013).

JH also supports reproduction indirectly, by balancing its trade-offs (Flatt et al., 2005). Met/Gce signalling suppresses immune defence in mated *Drosophila* (Schwenke and Lazzaro, 2017) or *A. aegypti* (Chang et al., 2021) females, maintains gut cell homeostasis necessary for increased metabolism in reproductive flies (Rahman et al., 2017; Reiff et al., 2015), adjusts carbohydrate and lipid metabolism (Wang et al., 2017a; Xu et al., 2013), and stimulates mating behaviour (Bilen et al., 2013) (Fig. 1). Some other JH functions such as differentiation of the termite soldier caste also require Met (Masuoka et al., 2015, 2018). Finally, it is interesting to note that the function of Met as a receptor for MF is conserved in crustaceans (Li et al., 2021a; Miyakawa et al., 2018; Qu et al., 2015).

3.2 Transcriptional regulation of JH-response genes

3.2.1 JHRs are bHLH-PAS protein complexes containing Met/Gce

Transcription factors of the bHLH-PAS family, to which Met belongs, form heterodimers of class I and class II members such as the circadian clock protein complex Clock-Bmal (the insect ortholog of Bmal is Cycle). The hypoxia inducible factor (HIF-α) or the aryl hydrocarbon receptor (AhR) dimerize with their common class II partner Arnt. The latter pair interacts when activated by low-molecular weight agonist ligands binding to the

PAS-B domain of AhR, upon which the complex recognizes specific DNA response elements to initiate transcription of target genes (Beischlag et al., 2008; Denison and Faber, 2017; Denison et al., 2011; Kewley et al., 2004).

By analogy to AhR, Met should also require another bHLH-PAS domain protein as a binding partner to form an active JHR complex. *Drosophila* Met was shown to interact with itself and with Gce in the absence of a JH agonist (Godlewski et al., 2006), but the exact nature of these homophilic complexes and their function remain unknown. Candidate members of the bHLH-PAS family from *A. aegypti* were tested for interaction with Met in yeast two-hybrid assays. This approach yielded two interesting candidates: the circadian clock protein Cycle (Shin et al., 2012) and an ortholog of *Drosophila* Taiman (Tai) (Li et al., 2011), which had been earlier described as a transcriptional coactivator of EcR (Bai et al., 2000; Zhu et al., 2006). Tai (also called FISC or SRC) is homologous to the vertebrate steroid receptor coactivator 1 (SRC-1). For clarity, and in keeping with a prevailing consensus in the current literature, we will further refer to this protein as Tai (Taiman) rather than SRC.

Reminiscent of AhR-Arnt dimer formation, presence of JH agonists induces association of Met with either Tai or Cycle. Mutations preventing Met from binding JH abolished the interaction with Tai but not formation of the homophilic Met-Met complex (Charles et al., 2011), indicating that the protein subunits of the active JHR assemble in an agonist-dependent manner. A similar effect of JH binding could be demonstrated for interaction between Gce/Met and Tai proteins from *Drosophila* and other insects using two-hybrid assays (Bittova et al., 2019; Miyakawa and Iguchi, 2017) (Fig. 4E and F).

The requirement for Tai in transducing the JH signal by Met/Gce has been corroborated by work on *A. aegypti* (Li et al., 2011, 2014; Liu et al., 2018), *T. castaneum* (Charles et al., 2011; Kayukawa et al., 2013; Zhang et al., 2011), *B. mori* (Kayukawa et al., 2012), *D. melanogaster* (Jindra et al., 2015b), *P. apterus* (Bajgar et al., 2013; Smykal et al., 2014a), *B. germanica* (Lozano et al., 2014), and *L. migratoria* (Guo et al., 2014; Wang et al., 2017b). Importantly, some of these studies have revealed the roles of Tai in insect metamorphosis and reproduction, both in holometabolous and hemimetabolous species (see also Jindra et al., 2015a; Roy et al., 2018, for reviews). Moreover, interactions between Tai and Met orthologs from cladoceran crustaceans, *Daphnia pulex* and *D. magna*, could be enhanced by JH agonists including MF, as shown using two-hybrid (Miyakawa and Iguchi, 2017; Miyakawa et al., 2013) and BRET (Kakaley et al., 2017) protein-protein interaction assays.

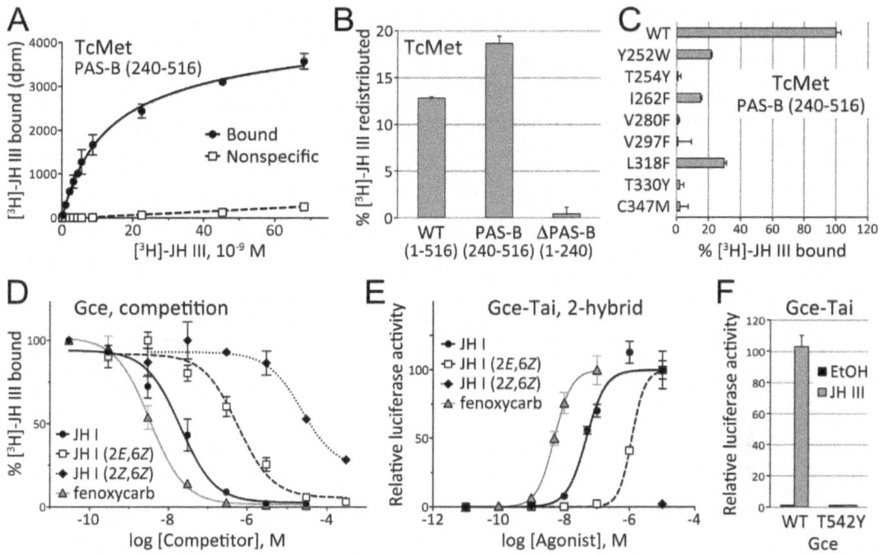

Fig. 4 Different methods demonstrate specific JH binding to the JHR proteins from *T. castaneum* (TcMet) and *D. melanogaster* (Gce). (A) Binding of [^{3}H]-JH III to the *in vitro* translated TcMet protein (the C-terminal half containing the PAS-B domain) as assessed using the dextran-coated charcoal method. A K_d value of 11.7 nM was determined in this experiment (performed by Jean-Philippe Charles). (B) Equilibrium dialysis with the entire (WT) or truncated *in vitro* translated TcMet led to a biased distribution of [^{3}H]-JH III towards those proteins containing the PAS-B domain. (C) Effects of mutations within the PAS-B domain of TcMet on the relative amount of [^{3}H]-JH III bound in the dextran-coated charcoal assay. (D) In the glass-fibre filter assay, JH I (but not its geometric isomers) efficiently competed ($K_i = 15$ nM) against [^{3}H]-JH III for binding to the *in vitro* translated Gce protein. Fenoxycarb outcompeted JH I. (E) A two-hybrid assay that measures the agonist binding-dependent interaction between VP16-Gce and Gal4-Tai fusion proteins expressed in the human HEK293 cells recapitulated the results shown in (D). (F) Mutated Gce (T542Y) that is incapable of binding [^{3}H]-JH III was inactive in the two-hybrid assay. T542 in Gce corresponds to T254 in TcMet (panel C). Data in (A–C) are from Charles et al. (2011); data in (D and E) are from Bittova et al. (2019). Data in all panels are mean values; error bars represent standard deviations.

Together, ample evidence establishes Tai as a broadly conserved, functional component of JHR complexes containing agonist-bound Met or Gce. At the same time, it should be kept in mind that Met/Gce likely interacts with additional bHLH-PAS partners, such as Cycle or the insect Arnt ortholog Tango (Tgo), as has been indicated by yeast two-hybrid assays with the *A. aegypti* proteins (Shin et al., 2012). The latter interaction did not require JH. Similarly, Met and Tgo from *T. castaneum* co-immunoprecipitated in a JH-independent manner (Jindra and Rynes, unpublished data).

3.2.2 Immediate target genes are driven by JH-response elements

As a JH receptor, Met should activate JH-inducible genes in an agonist-dependent manner. A number of JH-response genes became known through their involvement in processes regulated by JH. Ectopic induction of *Krüppel-homologue 1* (*Kr-h1*) by exogenous JH agonists in *Drosophila* and *Tribolium* pupae interfered with adult development (Minakuchi et al., 2008b, 2009), and JH-induced *Kr-h1* also blocked metamorphosis in last-instar hemimetabolous larvae (Konopova et al., 2011; Lozano and Bellés, 2011). These and additional studies in *Bombyx* (Kayukawa et al., 2012, 2014, 2017) have established Kr-h1 as a JH-inducible repressor of metamorphosis, universal to hemi- and holometabolous insects (Bellés, 2020; Jindra, 2019; Truman, 2019; Ureña et al., 2016). RNAi experiments initially revealed that Met was required for JH to induce *Kr-h1* mRNA expression (Konopova et al., 2011; Minakuchi et al., 2009; Zhu et al., 2010). Cell-based reporter studies complemented with chromatin immunoprecipitation or electrophoretic mobility shift assays (EMSA) confirmed that *Kr-h1* is indeed transcriptionally activated by binding of Met to its regulatory DNA elements (Cui et al., 2014; He et al., 2014; Kayukawa et al., 2012, 2013; Li et al., 2011, 2014; Zhang et al., 2011). Some of these studies have also implicated Tai as a component of the activating JHR complex. The *early trypsin* (*ET*) gene from *A. aegypti* (Noriega et al., 1997; Zhu et al., 2010) is another JH-inducible target, dependent on immediate transcriptional activation by the Met-Tai complex (Li et al., 2011, 2014).

The *ET* and *Kr-h1* genes provided material for detailed analyses of *cis*-acting JH-response elements (JHREs). A JHRE1 from the *A. aegypti ET* gene contained an imperfect palindrome (CACGCG) resembling *E*-box motifs bound by some bHLH-PAS proteins. JHRE1 was occupied by Met and Tai, which activated a JHRE1-driven luciferase reporter in response to JH III (Li et al., 2011). A similar reporter was induced by JH agonists through the endogenous Gce and Tai proteins in the *Drosophila* S2 cell line (Jindra et al., 2015b). JHREs from the *Kr-h1* genes (*k*JHREs) of several insects were described following systematic mutagenesis of regulatory sequences of the *Bombyx* and *Tribolium Kr-h1* orthologs (Kayukawa et al., 2012, 2013). These studies led to identification of a functional *k*JHRE with an *E*-box palindrome CACGTG. An unbiased selection of oligonucleotides bound by the complex of recombinant *Aedes* Met and Tai proteins essentially confirmed preference of the mosquito JHR for JHREs containing the core palindrome sequence (Li et al., 2014). A double-stranded DNA probe carrying this sequence bound the Met-Tai complex with high affinity ($K_d = 5.8\,nM$); K_d values of 13.8 nM and 103.0 nM were determined for the *Bombyx* *k*JHRE and the *Aedes* JHRE1, respectively (Li et al., 2014). By mutating N-terminal basic regions in Met and Tai, Li et al. (2014) have established that the bHLH domains of both interacting proteins are required for the DNA-binding capacity of the JHR complex. This situation resembles other bHLH-PAS transcription factors where each of the dimerizing partners contributes to form a bipartite DNA-binding domain (Kewley et al., 2004).

3.3 Binding JHs as agonist ligands

3.3.1 Validation of JH binding through independent assays

High-affinity, saturable and competitive binding of a cognate hormonal agonist is the key prerequisite for a receptor function. The hydrophobic, "sticky" nature and chemical instability of the native insect JHs posed many problems leading to false discoveries and artefacts (for a review see Riddiford, 2020). Therefore, having a reliable and properly controlled ligand-binding assay, validated by alternative methods based on different principles, was imperative. Tritium-labelled JH III ([10-^{3}H(N)]-JH III), formerly commercially available as a racemic mixture of 10*R* and 10*S* enantiomers, enabled several types of ligand-binding assays (Fig. 4).

A simple method, previously employed for hemolymph JH-binding proteins (JHBPs) (Touhara et al., 1993), relies on the capacity of dextran-coated charcoal to capture free, but not protein-bound [^{3}H]-JH III. Using this assay, Miura et al. (2005) initially reported a 5.3 nM K_d value for binding of [^{3}H]-JH III to the *Drosophila* Met protein, which was *in vitro* translated in rabbit reticulocyte lysates. Their experiment was difficult to repeat whether *Drosophila* Met was translated *in vitro* (Charles et al., 2011) or expressed using baculovirus in insect *Sf9* cells (M. Jindra, unpublished data). In contrast, the *Tribolium* Met and *Drosophila* Gce proteins proved high-affinity JH III binders, with K_d of 2.9 nM and 19.3 nM, respectively (Charles et al., 2011; Jindra et al., 2015b) (Fig. 4A–D). The ortholog of Met from the primitive ametabolous firebrat, *Thermobia domestica* (Zygentoma), also bound JH III avidly (Charles et al., 2011), indicating that the JHR function of Met predates the origin of insect metamorphosis. High-affinity binding to JH III ($K_d = 4.4\,nM$) was further reported for Met from *A. aegypti* (Li et al., 2014). While the above cited reports utilized proteins translated *in vitro*, specific JH III binding has been achieved with recombinant *T. castaneum* and *A. aegypti* JHR proteins expressed in and purified from the *Sf9* cells (Jindra et al., 2021). It should be noted that obtaining

JH-binding data for Met proteins from some insect species was unsuccessful, possibly due to improper folding of the intricate bHLH-PAS domain structure.

Importantly, the JH-binding capacity of Met/Gce could be validated through additional, independent methods. During equilibrium dialysis, [^{3}H]-JH III accumulated in the compartment of the dialysis chamber which contained an intact *Tribolium* Met protein or its ligand-binding domain, but not a truncated Met lacking this domain (Charles et al., 2011) (Fig. 4B). Another method was adapted from an early protocol for ecdysteroid binding studies (Cherbas et al., 1988; Graham et al., 2007). In this assay, receptor protein is captured on a glass-fibre filter, while any unbound ligand is washed away. Like the charcoal method, this filter assay enabled verification of specific binding for Gce or Met proteins (Bittova et al., 2019) (Fig. 4D). Finally, the ligand-dependent interaction of Met/Gce with Tai in a two-hybrid assay (Miyakawa and Iguchi, 2017) could be used as a proxy to the direct ligand-binding assays (Fig. 4E). Critical controls in any of the above experiments included Met/Gce proteins where the ligand-binding PAS-B domain was removed or mutated (Fig. 4B, C and F).

3.3.2 Ligand selectivity and stereoselectivity

Besides JH III and its precursor MF, common to most species, some insects produce additional types of JHs that differ by ethyl *vs.* methyl branches and by the number and position of epoxide groups (Goodman and Cusson, 2012; Hiruma and Kaneko, 2013; Tsang et al., 2020). A cell-based, JHRE-luciferase reporter assay indicated a slight preference (a 3.6-fold difference in ED_{50} values) of *Bombyx* Met2 towards the lepidopteran, ethyl-branched JH I relative to JH III (Kayukawa and Shinoda, 2015). This difference was cancelled when the Met PAS-B domain from *Tribolium* (whose native hormone is JH III) was substituted for that in *Bombyx* Met2. In addition to JH III, MF and JH III bisepoxide (JHB3) are circulating JHs in *Drosophila* (Dubrovsky and Bernardo, 2014; Richard et al., 1989; Wen et al., 2015). *In vitro* binding assays, based on competition of an unlabelled compound against [^{3}H]-JH III, have revealed that the Gce protein binds all of the three native *Drosophila* JHs, while showing no appreciable preference for JH III over the lepidopteran JH I (Bittova et al., 2019). Relative to the epoxidated JHs, the affinity of MF was about 8-fold lower (K_i values 90 nM and 11 nM for MF and JH III, respectively), indicating the relevance of the epoxide moiety for the bond with Gce. These results were corroborated for endogenous Gce using a JHRE-luciferase reporter in

Drosophila S2 cells (Bittova et al., 2019). Complementary data for *Drosophila* Met were obtained in a similar assay based in the transfected human HEK293T cell line, where Met showed a minor preference for JH III over JH I, and was slightly less activated by MF (Yokoi et al., 2020).

The study by Bittova et al. (2019) has also shown that Gce prefers the native (10*R*) enantiomer of JH III over its unnatural (10*S*) antipode, and that ligand binding by Gce is highly sensitive to the correct geometric isomerism of the JH skeleton (Fig. 4D and E). Testing all possible JH I isomers demonstrated that any deviation from the native 2*E*,6*E* conformation dramatically reduces binding to Gce and its agonist-dependent interaction with Tai, the transcriptional activation of *Kr-h1*, and the effect on *Drosophila* adult development. The 2*Z*,6*Z* conformation renders JH I inactive (Bittova et al., 2019) (Fig. 4D and E). The stereoselectivity of Gce corresponds to early data on JH isomer activities in bioassays (Wigglesworth, 1969) and matches binding affinities towards hemolymph JHBPs (Goodman et al., 1978).

3.3.3 Delineating the JH-binding pocket

Ligand-binding assays with truncated versions of *Tribolium* Met have localized JH-binding activity to the PAS-B domain (Charles et al., 2011) (Fig. 4A–C), corresponding to the site of ligand interaction in AhR (Pandini et al., 2007, 2009; Soshilov and Denison, 2014). Within their respective receptor complexes, AhR and Met/Gce are the ligand-sensing subunits. Thus, for instance, the interaction of *Tribolium* Met with Tai is induced by JH agonist binding to the PAS-B domain of Met, whereas Tai itself neither binds JH nor is required for ligand binding by Met (Charles et al., 2011).

While the PAS-B domain of AhR has so far escaped structural resolution, models for PAS-B domains of Met (Charles et al., 2011; Li et al., 2014) and Gce (Bittova et al., 2019) could be developed based on partial homology to HIF-2α, whose NMR and crystal PAS-B structures are available (Erbel et al., 2003; Scheuermann et al., 2009). The modelling and ligand docking images highlighted amino acid residues lining a putative JH-binding pocket. When individually replaced with amino acids of similar physicochemical nature but bulkier side chains, these mutations either reduced or completely abolished the capacity of *Tribolium* Met to bind [^{3}H]-JH III (Charles et al., 2011) (Fig. 4C). Molecular modelling coupled with mutagenesis and ligand-binding assays were also performed with the *Aedes* Met protein (Li et al., 2014), essentially confirming a common set of about 10 highly conserved residues required for binding JH.

Corresponding mutations prevented both JH binding and transcriptional activation by *Drosophila* Gce (Jindra et al., 2015b). A docking simulation based on the model of Gce PAS-B predicted a hydrogen bond between the epoxide oxygen in JH III and the hydroxyl group of a conserved tyrosine (Bittova et al., 2019). Removal of this tyrosine hydroxyl by mutation to phenylalanine substantially reduced [^{3}H]-JH III binding by Gce.

Besides probing the JH-binding pocket, the mutagenesis of Met/Gce has provided essential experimental means to decide whether any tested effect depends on JH binding. For instance, Met or Gce variants incapable of binding JH could not engage in the JH-induced interaction with Tai, either in co-immunoprecipitation (Charles et al., 2011) or two-hybrid (Fig. 4F) assays, and mutated Gce or Met could not activate a JHRE-driven reporter in a cell-based assay (Jindra et al., 2015b).

3.3.4 JH binding by Met/Gce is required for insect development

While the studies cited thus far have established the necessity of the receptor-hormone interaction in cell-based assays, it remained to be seen whether the JH-binding capacity of Met/Gce was also relevant *in vivo*. This could be tested in *Drosophila* deficient for Met and Gce (Jindra et al., 2015b). *Met* mutant flies are viable but abnormally resistant to methoprene and other JH agonists (Wilson and Ashok, 1998; Wilson and Fabian, 1986), whereas hemizygous *Met gce* double-mutant males never survive past pupation (Abdou et al., 2011). Ubiquitous transgenic expression of either Met or Gce rendered the *Met* mutants naturally sensitive (or hypersensitive in the case of Gce) to methoprene or pyriproxyfen treatment as long as the transgenic Met/Gce retained the JH-binding capacity (Jindra et al., 2015b). Accordingly, expressing functional Met or Gce was sufficient to rescue full viability in the unconditionally lethal *Met gce* males (Fig. 5). Mutations of individual amino acids that are critical for binding JH rendered both Met and Gce incapable of restoring the sensitivity to JH agonists and of sustaining the development of the fly. This genetic evidence has unequivocally proven that JH-binding capacity is essential for both Met and Gce to function as JH receptors *in vivo*.

3.3.5 Insect Met and Tai confer JH responsiveness to mammalian cells

A number of cell-based reporter assays for monitoring JH activity have been devised in both insect and mammalian cell lines (reviewed by Ito-Harashima and Yagi, 2021). While activation of JHRE-luciferase reporters in insect cells can rely on endogenous Met/Gce and Tai proteins (Bittova et al., 2019;

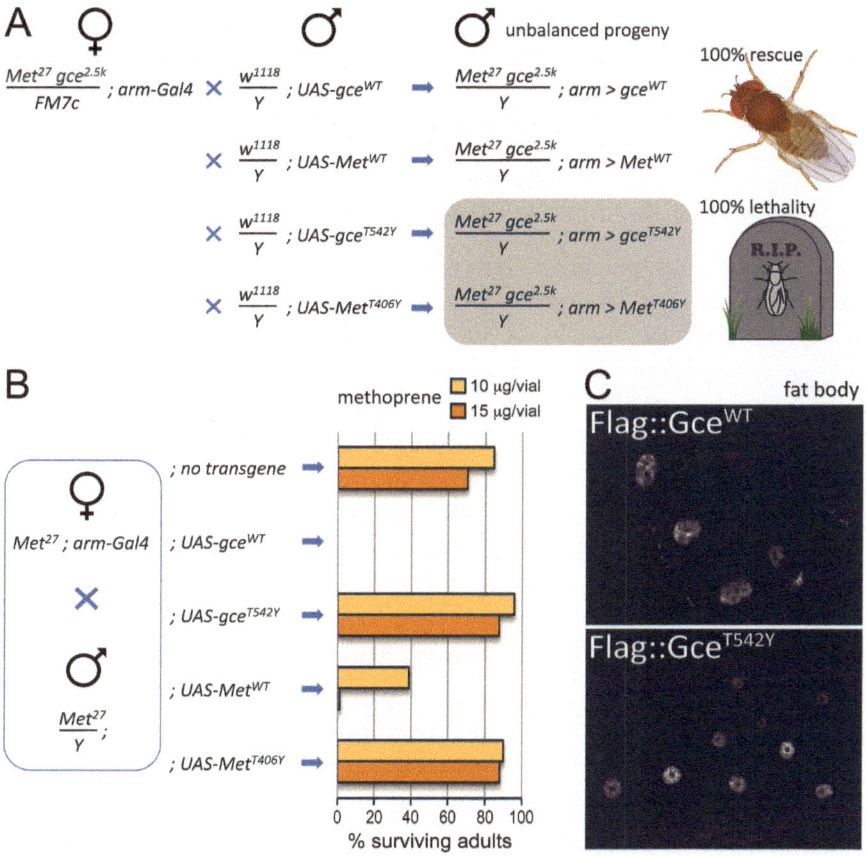

Fig. 5 Binding of JH is required for the function of the *Drosophila* Gce and Met JHR proteins *in vivo*. (A) The unconditionally lethal *Met*^{27} *gce*$^{2.5k}$ double-hemizygous males can be rescued to adulthood by ubiquitous (*armadillo*; *arm-Gal4*-driven) transgenic expression of either Gce or Met. *Met*^{27} *gce*$^{2.5k}$ males cannot complete development when the mutated Gce^{T542Y} or Met^{T406Y} variants incapable of binding JH are expressed instead. (B) *Met*^{27} flies are viable but resistant to methoprene. Transgenic expression of either Gce or Met but not of their mutated versions restore the natural methoprene sensitivity in the *Met*^{27} background. (C) While deficient in JH binding, the Flag epitope-tagged Gce^{T542Y} protein resides in the nuclei of the larval fat body as assessed by immunostaining (performed by Mirka Uhlirova). For details see Jindra et al. (2015b).

Jindra et al., 2015b; Kayukawa et al., 2012, 2013; Li et al., 2011), mammalian cells lack an ortholog of Met and thus are nonresponsive to JH. However, strong expression of JHRE-driven reporters could be induced by JH agonists in human (HEK293) cells transfected with constructs encoding *Bombyx* or *Drosophila* Met and Tai proteins (Kayukawa and Shinoda, 2015;

Kayukawa et al., 2012, 2021; Yokoi et al., 2020). Transfection with either Met or Tai alone was insufficient for the reporter activation. These studies have demonstrated that both Met and Tai are required and sufficient to constitute an active JHR complex in heterologous cells.

4. Molecular architecture and function of JHR

Intracellular receptors comprise ligand-binding and gene *trans*-activation functions. In order to regulate target genes, such receptors translocate to cell nuclei, recognize specific *cis*-regulatory DNA elements, and engage in contacts with transcriptional coactivators or corepressors. Moreover, their DNA-bound, active forms are typically dimers of the same or different members of the given protein family. These multiple functions are reflected in the modular architecture of the receptor molecules. In this regard, nuclear receptors of the NR family and the bHLH-PAS receptors such as AhR and Met/Gce are functionally analogous in spite of their unrelated molecular structures.

4.1 Functional parallels between Met/Gce and AhR

AhR exemplifies a bHLH-PAS transcription factor that is activated by agonist ligands. While AhR is neither a hormonal receptor nor a true ortholog of Met/Gce, it is the closest vertebrate homologue by sequence, and probably the best available proxy in terms of mode of action. Considering AhR function may thus be a useful guide to begin to understand how JHR works. As the Met and AhR proteins share a common functional domain architecture (Fig. 6A), some parallel mechanisms emerge.

4.1.1 The canonical AhR pathway

The mass of knowledge on AhR signalling has been covered in several comprehensive reviews (*e.g.*, Beischlag et al., 2008; Denison and Faber, 2017; Denison and Nagy, 2003; Denison et al., 2011; Jackson et al., 2015; Stejskalova et al., 2011; Stockinger et al., 2014). In the absence of an agonist [such as the environmental pollutant 2,3,7,8-tetrachlorodibenzo-*p*-dioxin (TCDD)], inactive AhR resides in the cytoplasm, bound in a multiprotein complex including the chaperone heat-shock protein 90 (Hsp90), a co-chaperone p23, and an AhR interacting protein (AIP/XAP2). Exposure to TCDD and its high-affinity binding to the PAS-B domain of AhR triggers a conformational change that allows the AhR complex to translocate *via* β-importin to the cell nucleus, where AhR dissociates from Hsp90

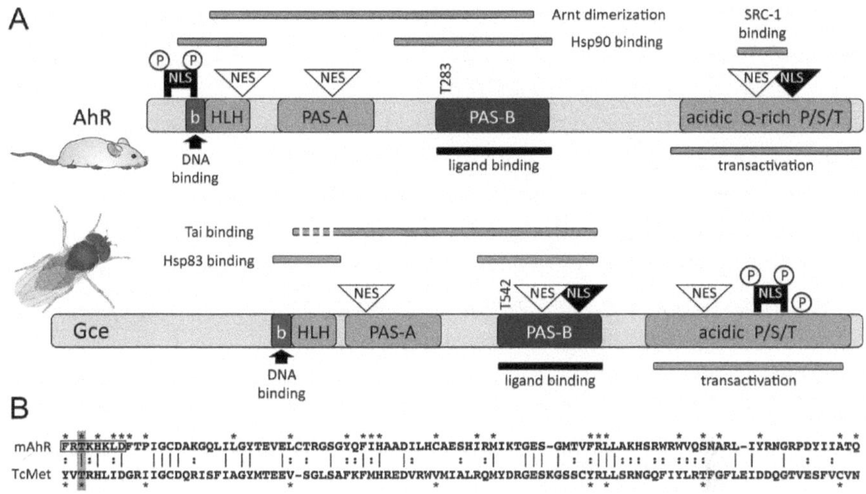

Fig. 6 Analogy of functional domains in mouse AhR and *Drosophila* Gce proteins. (A) The approximate positions of functional domains are denoted (b, basic; HLH, helix-loop-helix; PAS, Per-Arnt-Sim; P/S/T, proline/serine/threonine-rich). Interaction interfaces are indicated below and above the schemes of each protein. NES (nuclear export signal) and NLS (nuclear localization signal) are indicated by triangles or, for bipartite NLS, by black rectangles. Circled Ps denote areas of phosphorylation. Threonine residues critical for ligand binding are indicated. (B) The ligand-binding pockets of the mouse AhR (mAhR) and *T. castaneum* Met (TcMet) show similarity in some positions (asterisks) that are involved in binding of agonists (TCDD and JH, respectively). Strand A of the putative β-sheet in mAhR (open box) is critical for contacting Hsp90. The shaded box indicates the conserved residues T283 (mAhR) and T254 (TcMet, corresponding to T542 in Gce) that are required for ligand binding.

(Soshilov and Denison, 2011; Soshilov et al., 2020). The liganded AhR dimerizes with Arnt and the resulting complex binds enhancers such as the xenobiotic/dioxin response elements to induce transcription of detoxification genes including CYP1A1. There are numerous deviations from this canonical xenobiotic response. These entail diverse AhR ligands (including endogenous agonists) and different protein partners [*e.g.*, the oestrogen receptor (Ohtake et al., 2003, 2007)], enabling AhR to regulate alternative target genes and hence many relevant cellular responses (Denison and Faber, 2017; Denison and Nagy, 2003; Jackson et al., 2015; Stejskalova et al., 2011; Stockinger et al., 2014).

4.1.2 Interactions with bHLH-PAS partners

Crystal structures capturing the bHLH and PAS-A domains of the AhR-Arnt dimer in a complex with the dioxin response element have determined the contact interfaces between the two proteins and DNA at atomic resolution (Schulte et al., 2017; Seok et al., 2017). This work has revealed the individual amino acids responsible for the AhR-Arnt dimerization and DNA binding within the bHLH and PAS-A domains. While the PAS-B structure of AhR has not yet been resolved, the amino acid residues effecting contacts between the PAS-B domains of AhR and Arnt could be identified through modelling and functional assays with the proteins mutated at specific sites (Corrada et al., 2016, 2017; Soshilov et al., 2020).

Similar to AhR, Met must interact with another bHLH-PAS protein, such as Tai, in order to activate transcription (Kayukawa et al., 2012; Li et al., 2011; Zhang et al., 2011), and the basic motifs within the bHLH of both Met and Tai are required for binding the JHRE DNA (Li et al., 2014). The regions of mutual interaction have been mapped, albeit only roughly, for the *Aedes* Met-Tai complex (Li et al., 2011) (Fig. 6A). While both PAS-A and PAS-B domains of Met were essential for JH-induced interaction with Tai in a two-hybrid assay, the bHLH domain of Met seemed dispensable. However, this does not rule out a participation of the bHLH domain in forming the protein complex in a manner similar to the bHLH domain of AhR. Conversely, deletion of the PAS-B and, to a lesser extent, of the PAS-A and bHLH domains from *Aedes* Tai reduced JH-dependent interaction with Met (Li et al., 2011). Co-immunoprecipitation of *Tribolium* Met and Tai demonstrated that the PAS-B domain of Met was both necessary and sufficient for JH-induced binding to Tai (Charles et al., 2011).

It is worth noting that Tai is neither an insect ortholog of Arnt nor a typical class II bHLH-PAS protein. SRC-1, the mammalian ortholog of Tai, directly contacts AhR to enhance its transcriptional activity in response to agonists such as TCDD (Beischlag et al., 2002; Zhang et al., 2008). However, rather than acting through the PAS domains, LXXLL motifs in the central region of SRC-1 interact with the glutamine-rich C-terminal transactivation part of AhR (Kumar and Perdew, 1999) (Fig. 6A). Therefore, the architecture of the JHR complex consisting of Met and Tai may differ from that of the AhR-SRC-1 and AhR-Arnt heterodimers.

4.1.3 The role of Hsp90/Hsp83

Interaction with the chaperone Hsp90 (orthologous to Hsp83 in insects) is a common feature of AhR and Met action (Fig. 6A). The role of Hsp90 is not only to keep the unliganded AhR inactive while in the cytoplasm, but also to protect AhR from proteasomal degradation. Furthermore, Hsp90 is critical for conferring competence of AhR to bind agonist ligands to activate transcription (Beischlag et al., 2008; Lees and Whitelaw, 1999; Pongratz et al., 1992; Whitelaw et al., 1995). A similar requirement applies to the ligand-dependent activities of some NRs such as the glucocorticoid receptor (Picard, 2006); EcR requires Hsp83 to bind DNA (Arbeitman and Hogness, 2000). Moreover, the chaperone complex including Hsp90 facilitates nuclear import of AhR (Kazlauskas et al., 2001). Hsp90 contacts AhR at the bHLH and PAS-B domains, of which the latter encompasses the ligand-binding site and partly overlaps with the dimerization interface for Arnt (Soshilov and Denison, 2011) (Fig. 6A). Simultaneous interactions of liganded AhR with Hsp90 and Arnt therefore appear to be mutually exclusive or, rather, limited to a brief transitional state (Soshilov et al., 2020). Whilst the PAS-B domain mediates activation by agonist ligands, its interaction with Hsp90 also blocks AhR in the absence of an agonist. Therefore, under *in vitro* conditions, deletion of the PAS-B domain lifts inhibition by Hsp90 and turns the truncated AhR into a constitutive DNA binder independent of TCDD (Soshilov and Denison, 2008).

The Hsp90-AhR mechanism is at least partially conserved for Hsp83 and Met in insects (He et al., 2014; Liu et al., 2013). In *Drosophila*, Hsp83 has been implicated in JH agonist-stimulated nuclear import of Met and induction of the JH-response gene *Kr-h1*. Loss of Hsp83 function or its chemical inhibition with geldanamycin in *Drosophila* larvae blocked the JH response. Short deletions within the *Drosophila* Met protein revealed that, like in AhR, the bHLH and PAS-B domains of Met were involved in the interaction with Hsp83 (He et al., 2014) (Fig. 6A). Nuclear import of the Met-Hsp83 complex was further shown to depend on nucleoporin Nup358 (He et al., 2017), a nuclear pore protein associated with β-importin. Both Hsp83 and Nup358 were captured through affinity binding to a JH-responsive upstream DNA sequence of *Kr-h1* among proteins extracted from *Drosophila* cells that had been cultured with methoprene (He et al., 2014). This is somewhat surprising, given that agonist binding to the PAS-B domain displaces Hsp90 from the complex with AhR (Soshilov and Denison, 2011) and that Hsp90 is unable to bind AhR once engaged in the active AhR-Arnt heterodimer (Beischlag et al., 2008).

4.1.4 Ligand-binding pockets and rich agonist repertoires

The ability to bind a diverse range of activating ligands is perhaps the most intriguing and distinguishing common feature of AhR and Met/Gce. As the three-dimensional structure of the PAS-B in AhR has not yet been resolved, our understanding of the receptor-agonist interactions has thus far relied on molecular modelling, supported by site-directed mutagenesis and testing of the mutated protein variants in ligand- and DNA-binding assays (Tagliabue et al., 2019; Motto et al., 2011; Pandini et al., 2007, 2009; Soshilov and Denison, 2014; Soshilov et al., 2020). Structures of HIF-α proteins in complexes with Arnt and/or artificial low-molecular ligands have been employed as the closest available homology templates for modelling AhR. Interestingly, some of the artificial ligands co-crystallized with HIF-2α proved to be competitive agonist ligands of AhR, further strengthening the relevance of the homology template (Motto et al., 2011; Tagliabue et al., 2019). Crystallographic and NMR data for HIF-2α have also been applied towards initial characterization of the JH-binding pocket in Met/Gce (Bittova et al., 2019; Charles et al., 2011; Li et al., 2014) (see Section 3.3.3) although the *Drosophila* circadian clock protein Period could serve as an alternative template (Bernardo and Dubrovsky, 2012; Yokoi et al., 2021).

Analyses of AhR have identified specific amino acid residues in the PAS-B domain whose mutations result in a partial or complete loss of TCDD- or other ligand-binding activity (Pandini et al., 2007, 2009; Soshilov and Denison, 2014; Soshilov et al., 2020; Tagliabue et al., 2019). These include a stretch of amino acids forming the A strand of the PAS-B β-sheet, known to mediate contact with Hsp90 (Soshilov and Denison, 2011; Soshilov et al., 2020). This short sequence is well conserved in insect Met/Gce proteins (Fig. 6B), including an invariable threonine that is required for [^{3}H]-JH III binding by *Drosophila* Gce and *Tribolium* and *Aedes* Met proteins (Charles et al., 2011; Jindra et al., 2015b; Li et al., 2014) (Fig. 4C). Additional point mutations within the PAS-B domain of Met/Gce that disable JH III binding occur at positions known to affect TCDD binding by AhR and often represented by amino acids with similar properties (Fig. 6B). These positions are also conserved in crustacean (*Daphnia*) Met orthologs and likewise predicted to contact JH agonists (Hirano et al., 2020).

Both AhR (Denison and Faber, 2017; Denison and Nagy, 2003; Denison et al., 2011; Soshilov and Denison, 2014; Stejskalova et al., 2011) and Met/Gce (Bittova et al., 2019; Charles et al., 2011; Jindra et al., 2015b) display a remarkable capacity to bind ligands of diverse chemistry. Examples for Met/Gce include insecticidal JH mimics (juvenoid IGRs) (Jindra and Bittova, 2020). It seems paradoxical that while Gce exhibits strict stereoselectivity towards the native hormones (Bittova et al., 2019) (Fig. 4D), it can accommodate many structurally unrelated agonists. Not surprisingly, disparate ligands could be docked to the computational models of Met/Gce PAS-B domains (Charles et al., 2011; Hirano et al., 2020). However, being based on the rather distant homology to the *apo* structure of HIF-2α, those rigid models contained large pockets, allowing room for diverse potential ligands.

Advanced modelling of AhR builds on multiple HIF-2α structures crystallized alone or with various artificial ligands, thus affording flexibility in shape and size of the ligand-binding cavity (Tagliabue et al., 2019). Ten AhR agonist ligands representing diverse chemistries were docked to the different models and results were reconciled with their actual binding affinities. This approach has unveiled some ligand-specific interactions within the ligand-binding pocket of AhR, suggesting a mechanism underlying the broad agonist spectrum.

4.2 Nuclear import

When expressed in insect tissues or heterologous cell lines, the Met and Gce proteins are predominantly nuclear or, to a variable degree, dispersed through both nuclei and the cytoplasm (Greb-Markiewicz et al., 2011, 2015; He et al., 2014, 2017; Jindra et al., 2015b; Miura et al., 2005; Pursley et al., 2000) (Figs. 5C and 7). Nuclear localization is enhanced by JH agonist treatment or when JH is naturally abundant *in vivo*. As already explained above, ligand-stimulated nuclear enrichment of *Drosophila* Met requires Hsp83 and importin machinery (He et al., 2014, 2017) (Fig. 8), similar to nuclear import of AhR.

AhR is subject to nucleocytoplasmic shuttling dependent on importin and leptomycin B-sensitive exportin/CRM1 activities (Ikuta et al., 1998, 2000, 2004). Intracellular trafficking of AhR relies on nuclear localization and export signals (NLS and NES, respectively), which were primarily identified in the N-terminal sequence of AhR (Ikuta et al., 1998) (Fig. 6A). A bipartite NLS consisting of two short stretches of basic amino acids overlaps with the bHLH domain and is masked by Hsp90 in the unliganded state. Agonist binding to AhR exposes the NLS, permitting nuclear import of the AhR-Hsp90 complex. Phosphorylation or phosphomimetic mutation of two serine residues immediately abutting the basic NLS sequence inhibits nuclear import of AhR (Ikuta et al., 2004). The N-terminal NES overlaps with the Arnt-binding interface and its blocking by Arnt or by phosphorylation of a serine residue within the NES retains AhR in the nucleus. Additional sequences that regulate intracellular trafficking of AhR have been identified in the PAS-A domain (Berg and Pongratz, 2001) and in the C-terminal region rich in Gln (Q) and Pro/Ser/Thr (P/S/T) residues (Tkachenko et al., 2016) (Fig. 6A).

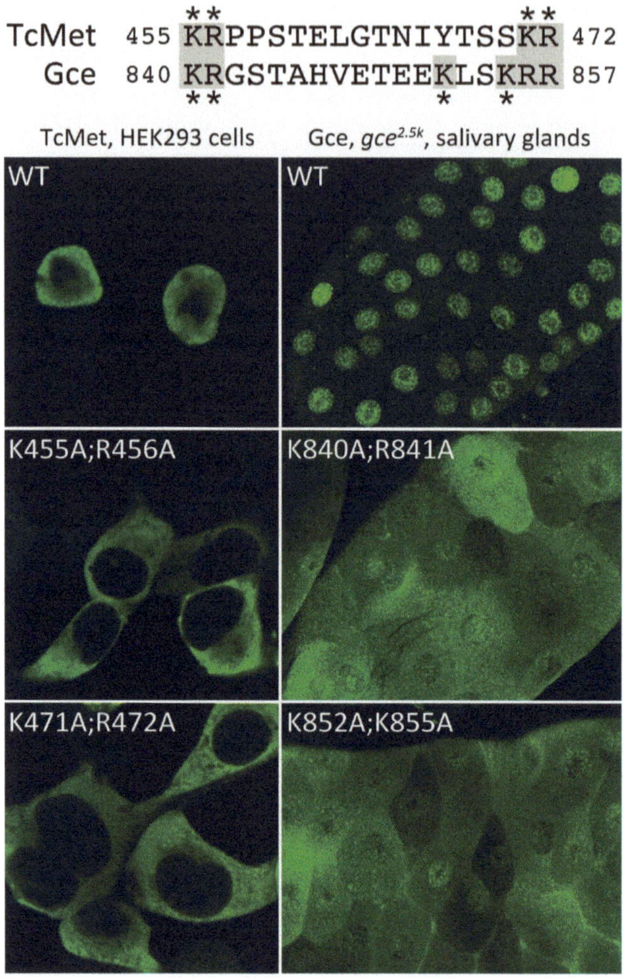

Fig. 7 Conservation of a bipartite NLS in the carboxyl termini of *T. castaneum* Met (TcMet) and *Drosophila* Gce proteins. The asterisks indicate basic amino acids in each protein that were mutated to alanine. The Myc epitope-tagged wild type (WT) and mutated TcMet proteins were expressed in human HEK293 cells; the variants of Gce were expressed from transgenic constructs in the salivary glands of homozygous $gce^{2.5k}$ wandering third-instar larvae. As detected by immunocytochemistry, the WT proteins are predominantly nuclear whereas the mutated versions occur in the cytoplasm. Images are single confocal slices.

Functional NLS and NES motifs have been found in the *Drosophila* Met and Gce JHR proteins, expressed with N-terminal YFP-tags in mammalian cell lines (Greb-Markiewicz et al., 2011, 2015). By means of systematic deletions, the authors delimited an NES in each of the PAS-A, PAS—B, and the

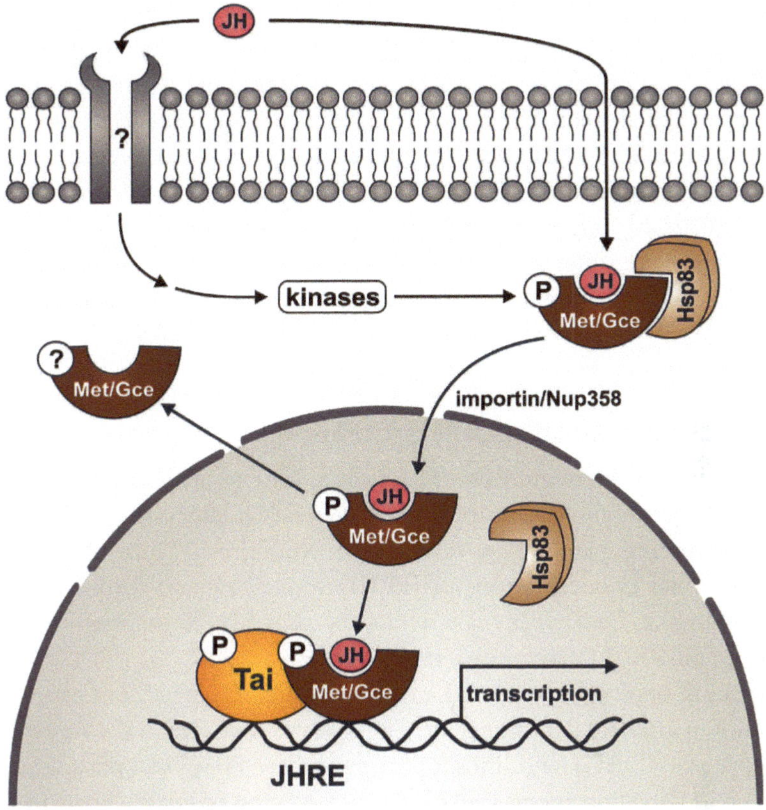

Fig. 8 A scheme of the gene-regulatory JHR action. In addition to regulation by JH and Hsp83, Met/Gce is also phosphorylated, at least partly as a result of JH acting through a hypothetical receptor at the cell membrane. Whether and how phosphorylation affects nucleocytoplasmic shuttling of Met/Gce and its interaction with Hsp83 is presently unknown.

C-terminal regions of either protein (Fig. 6A). Both Met and Gce also contain two NLS motifs, which reside either in each of the PAS domains of Met or in the PAS-B domain and near the carboxyl terminus of Gce. No NLS or NES have been predicted in N-terminal regions of *Drosophila* Met and Gce that would correspond to the positions of the signals in AhR (Greb-Markiewicz and Kolonko, 2019).

The NLS located in the PAS-A of *Drosophila* Met is a JH-independent signal for constitutive nuclear localization which, however, is not conserved in Gce or in Met proteins from other insects. In contrast, the C-terminal NLS in *Drosophila* Gce (Fig. 6A) contributes to the JH-induced nuclear import (Greb-Markiewicz et al., 2015). While absent from *Drosophila* Met, this bipartite NLS is functionally conserved in the C-terminal region of *Tribolium* Met, with each cluster of basic residues being required for nuclear localization of the beetle protein (Jindra et al., 2021) (Fig. 7). The position of the bipartite NLS in carboxyl termini of Gce and *Tribolium* Met is intriguing, as these non-conserved regions are intrinsically disordered (Kolonko et al., 2016, 2020) and may thus be prone to post-translational modifications. Indeed, we have detected phosphorylation of serine and threonine residues straddling the bipartite NLS in the *Tribolium* Met protein (Jindra et al., 2021).

4.3 Phosphorylation of JHR proteins

As mentioned at the beginning of Section 3, JH exerts rapid effects initiated at the cell membrane in addition to its gene-regulatory function (Fig. 1). Besides inducing patency in the follicle epithelium (Jing et al., 2018), sub-nanomolar concentrations of JH III were shown to stimulate protein synthesis in the *Drosophila* male accessory glands in a manner dependent on Ca^{2+} and PKC (Yamamoto et al., 1988).

Working on the fat body and a cell line from *A. aegypti*, Liu et al. (2015) have shown that JH III triggers a Ca^{2+} wave, followed by a surge of the diacylglycerol (DAG) and inositol trisphosphate (IP_3) second messengers. These rapid signalling events lead to phosphorylation of intracellular JHR proteins Met and Tai, resulting in their increased capacity to occupy the JHREs of the *ET* and *Kr-h1* genes and to induce their transcription in response to JH. In other words, JH also appears to engage in alternative, membrane-initiated signalling pathways that potentiate gene-regulatory functions of its own intracellular receptor (Fig. 8). Based on treatments with chemical inhibitors and RNAi, this phosphorylation pathway may involve Ca^{2+}/calmodulin dependent kinase II (CaMKII) (Liu et al., 2015) and PKC (Ojani et al., 2016) acting downstream of phospholipase C (PLC). Inhibitor experiments have also led the authors to postulate a receptor tyrosine kinase (RTK) as a hypothetical cell-membrane receptor for JH (Liu et al., 2015). Although such a receptor is yet to be discovered, the work has added a potentially important rapid-response element to regulation of JH signalling (Fig. 8).

The data from Liu et al. (2015) suggest that several as yet unidentified positions in the *Aedes* Met and Tai proteins may be phosphorylated, some in a JH-dependent manner. Indeed, our phosphoproteomic analysis of recombinant *Tribolium* and *Aedes* JHR protein complexes, purified from the *Sf9* cells, detected multiple phosphorylated amino acid residues, some enriched following exposure of the cell culture to a JH agonist. Most of the phosphorylations occurred in the intrinsically disordered C-terminal regions of the Met proteins (Jindra et al., 2021) (Fig. 6A). A current study (Li et al., 2021b) has proposed that, surprisingly, a single threonine within the ligand-binding domain of the Met1 protein from a moth, *Helicoverpa armigera*, becomes phosphorylated in the presence of JH. This threonine phosphorylation prevents the homophilic Met1-Met1 association and instead promotes a productive interaction between Met1 and Tai, the binding of the complex to a JHRE DNA, and the transcriptional activation of the *Helicoverpa Kr-h1* gene (Li et al., 2021b). Whether this is a conserved mechanism remains to be investigated.

5. JHR as a target of insecticide action

JH mimics are attractive for insect control because they hijack the arthropod-specific endocrine system without affecting vertebrates or even non-target invertebrates. The latter possibility has been exemplified by bis-acylhydrazine ecdysone receptor agonists (Hill et al., 2012; Pener and Dhadialla, 2012). The potential of JH signalling for controlling insect pests and disease vectors was recognized very early (Williams, 1967). The idea was inspired by discovery of a terpene juvabione in paper produced from American fir trees. Juvabione proved extremely potent and selective towards *Pyrrhocoris* and its closest relatives (Sláma and Williams, 1965). The chemical structure of juvabione (Bowers et al., 1966) is quite distinct from structures of native insect JHs, the first of which (JH I) was reported soon thereafter (Röller et al., 1967). The first synthetic JH mimic methoprene (Henrick et al., 1973) was registered in 1975, almost four decades before its molecular target, Methoprene-tolerant (Wilson and Ashok, 1998; Wilson and Fabian, 1986), became established as a JHR (Charles et al., 2011). Methoprene and other aliphatic juvenoids (Henrick, 2007) were followed by carbamate and pyridine derivatives such as fenoxycarb and pyriproxyfen, respectively (Dorn et al., 1981; Hatakoshi et al., 1986), in which aromatic or heterocyclic rings replaced the isoprene units of JH (Fig. 9). These juvenoid insecticides of the IGR category are more stable than JH and are highly active against a broad spectrum of insects (Minakuchi and Riddiford, 2006; Parthasarathy et al., 2012; Pener and Dhadialla, 2012; Sláma et al., 1974).

5.1 Juvenoid IGRs are agonist ligands of JHR

Given the practical utility of the JH mimics as insecticides and as tool compounds substituting for native JHs in hundreds of research studies, it is critically important to know whether these compounds act through the same

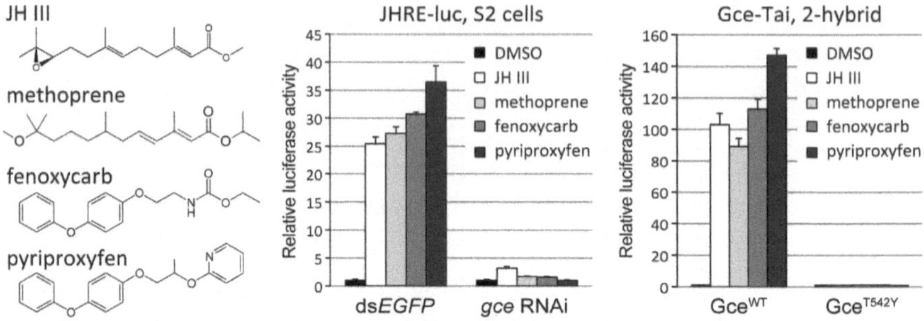

Fig. 9 JH and its synthetic agonists (juvenoid IGRs) act through the Gce protein in a ligand binding-dependent manner. All depicted compounds (left) activated a JHRE-luciferase reporter in the *Drosophila* S2 cells (centre). Activation was cancelled by knocking down *gce*, but not by treatment with a control (*EGFP*) dsRNA. When incapacitated for JH binding, Gce^{T542Y} could no longer engage in the agonist-dependent interaction with Tai. Data are mean values; error bars represent standard deviations.

molecular mechanism as the hormone itself. In spite of their diverse chemistries, the few synthetic JH mimics tested thus far competed against [^{3}H]-JH III for binding to the recombinant *Tribolium* and *Drosophila* Met and Gce proteins (Bittova et al., 2019; Charles et al., 2011; Jindra et al., 2015b). Moreover, as inferred from these competition assays, the binding affinities of the most potent IGRs such as pyriproxyfen or fenoxycarb were in the nanomolar range and surpassed those of JH III itself (Fig. 4D).

In vitro ligand-binding data could be complemented by two-hybrid assays wherein Met/Gce and Tai proteins, fused to the VP16 *trans*-activator and Gal4 DNA-binding domains, respectively, are expressed in mammalian cells (Miyakawa and Iguchi, 2017). As JH binding to VP16-Met/Gce stimulates its interaction with Gal4-Tai, the activity of a Gal4-driven luciferase reporter gauges the potency and efficacy of any tested agonist ligand. This two-hybrid assay clearly discriminated biologically active and inactive geometric isomers of JH I, and confirmed that relative to any native JH, fenoxycarb is a more potent agonist ligand of Gce (Bittova et al., 2019) (Fig. 4E). Importantly, when mutated at the threonine residue critical for binding JH III, VP16-Gce could no longer respond to synthetic JH mimics by dimerizing with Tai (Fig. 9). Fenoxycarb or pyriproxyfen also could not activate a JHRE-luciferase reporter in *Drosophila* S2 cells subjected to RNAi-mediated knockdown of either *gce* or *tai* (Bittova et al., 2019; Jindra et al., 2015b) (Fig. 9).

Prior knockdown of *Met* prevented the anti-metamorphic effects of JH mimics in *Tribolium* pupae (Konopova and Jindra, 2007, 2008; Minakuchi et al., 2009) and in last-instar larvae of *Pyrrhocoris* (Konopova et al., 2011). Similarly, adverse effects of pyriproxyfen on mosquito female reproduction and metabolism required Met in *Aedes* (Ahmed et al., 2020). Finally, unlike functional Met or Gce, the mutated proteins incapable of binding JH failed to restore the natural methoprene or pyriproxyfen sensitivity when expressed in the *Methoprene-tolerant* mutant flies (Jindra et al., 2015b) (Fig. 5B). Therefore, the JH-binding function is required for Met/Gce to mediate the developmental effect of the juvenoid IGRs *in vivo*.

Based on the evidence summarized above, it can be inferred that at least some synthetic JH mimics activate the intracellular JHR as genuine agonist ligands (Jindra and Bittova, 2020). Nonetheless, one common molecular mechanism for all JH mimics cannot be taken for granted. The chemical diversity of natural and synthetic compounds exerting JH-like effects is extensive, perhaps comparable to that of natural and man-made agonists of AhR. It is thus likely that some JH agonists interact with the JHR in distinct modes, as has been predicted for binding of diverse agonists to the ligand-binding pocket of AhR (Tagliabue et al., 2019). Ligand-specific modes of interaction have been demonstrated rigorously for the 20E receptor EcR through resolving crystal structures of EcR bound by various steroid or nonsteroidal agonists (Billas et al., 2003; Browning et al., 2021; Carmichael et al., 2005; Hill et al., 2012; Iwema et al., 2007; Ren et al., 2014). Some nonsteroidal agonists induced major conformational changes to the ligand-binding pocket, illustrating its adaptability to the diverse ligand chemistries (Browning et al., 2021).

5.2 Search for novel agonists and antagonists of JHR

While just a handful of juvenoid IGRs have been commercialized, a large number of compounds that exert JH-like effects on insects, reportedly over six thousand, have been produced (Parthasarathy et al., 2012). It remains to be ascertained whether all these compounds might be JHR agonists as has been shown for methoprene, fenoxycarb, and pyriproxyfen (Jindra and Bittova, 2020). In striking contrast, no authentic JHR antagonists have been obtained to date.

The currently available juvenoid IGRs have disadvantages such as low selectivity to target insect species and toxicity to crustaceans, while resistance and cross-resistance have also arisen in mosquitoes, whiteflies, lepidopterans, and other insects (reviewed by Pener and Dhadialla, 2012). Moreover, juvenoids prevent metamorphosis and prolong the larval feeding period, in some cases leading to giant supernumerary larvae and increased crop losses. A related problem is that JH agonists are often ineffective until late during development. In contrast, a JH antagonist would terminate the larval stage early and provoke a fatal precocious metamorphosis. Therefore, JH antagonists should in theory surpass JH mimics in insect control. They would also be very useful as tool compounds for research.

5.2.1 *Activators discovered* via *high-throughput and virtual screening*

Efforts are underway to discover novel compounds targeting the JH signalling pathway either positively or through specific inhibition. These efforts are greatly aided by the knowledge of the JHR. The structure of the JH-binding pocket in Met/Gce remains to be resolved before a rational design of novel compounds may be attempted. Meanwhile, cell-based platforms built on the agonist-dependent capacity of the JHR to induce JHRE-driven reporters can be exploited in high-throughput screening (HTS) of chemical libraries. This approach has been taken by Kayukawa et al. (2021). They selected a stable *Bombyx* cell line carrying a *k*JHRE-luciferase reporter to screen a diversity collection of 9600 compounds. The HTS yielded nine confirmed primary hits of which six proved dose-dependent activators of the *k*JHRE reporter. Compounds of analogous structures were then retrieved from a larger library, expanding the number to 10 validated activators, of which some shared structural features with existing IGRs such as pyriproxyfen, while others were distinct. Seven of the activators could delay metamorphosis of final-instar *Bombyx* larvae although none was able to block pupation and provoke an extra larval moult. Although it remains to be determined whether these JH signalling activators bind *Bombyx* Met as agonist ligands, they present lead compounds with a potential for further development (Kayukawa et al., 2021).

A chemical HTS is currently concluding in collaboration between our laboratory and Dr. Sedlák of the national core facility for chemical biology (CZ-OPENSCREEN). Starting from a large set of compounds, our campaign produced a great diversity of validated hits, corresponding to the structural repertoire of known JHR agonists. The outcome is somewhat reminiscent of the situation with AhR, for which an extensive chemical HTS has revealed an extreme ligand promiscuity (Denison, 2010).

Building pharmacophore models based on quantitative structure-activity relationship (QSAR) of juvenoid compounds active against *Drosophila* was initially attempted without considering a JHR protein (Liszekova et al., 2009; Parthasarathy et al., 2012). However, the first successful example of a JH agonist discovered through an *in silico* approach was recently set by Yokoi et al. (2021). Their virtual screen of five million compounds relied on a structural model of *Drosophila* Met. Virtual compounds selected based on similarity to pyriproxyfen and fenoxycarb were computationally docked to the model, finally yielding 11 hits. Candidate compounds were validated in a JHRE-driven reporter assay in human cells expressing the *Drosophila* Met and Tai proteins (Yokoi et al., 2020). One piperazin-based hit displayed a sub-micromolar potency, thus providing a novel lead compound (Yokoi et al., 2021). Its JHR agonist activities need to be tested in additional assays and *in vivo*.

One goal of screening for new agonists is to find compounds selective against target species. Juvabione (Bowers et al., 1966; Sláma and Williams, 1965) and certain small peptidic derivatives, such as ethyl L-isoleucyl-L-alanyl-*p*-aminobenzoate (Sláma et al., 1974; Zaoral and Slama, 1970), exemplify JH mimics displaying an exquisite selectivity towards true bugs of the Pyrrhocoridae family. Whether and how the action of such compounds reflects differences within the ligand-binding pockets of JHR proteins from specific insects is currently under investigation.

5.2.2 Inhibitors

Relative to agonists, JHR antagonists appear to be more challenging to find. Besides the absence of precedent validated antagonists, the problem may be that there are many ways in which any reporter assay can be inhibited. Nevertheless, it is encouraging to know that both allosteric and competitive antagonists exist for AhR (Denison and Faber, 2017; Denison and Nagy, 2003; Denison et al., 2011; Stejskalova et al., 2011).

One approach to discovering JHR inhibitors stems from the idea that plants might produce anti-JH compounds as a natural defence against insect herbivores. Using the yeast two-hybrid assay for agonist-stimulated interaction between Met and its partner proteins, diterpenic and acyclic alkyl compounds were extracted from plants (Lee et al., 2015, 2018; Oh et al., 2017; Shin et al., 2018). At $>10\,\mu M$ doses, some of the diterpenes interfered with assembly of JHR complexes, while some appeared to affect JH-dependent processes in mosquitoes and other insects. However, it is unclear whether these natural products directly antagonize the JHR or whether they compromise interaction of its subunit proteins in other ways.

The same chemical HTS in the *Bombyx* cell line described above for JHR agonists (Kayukawa et al., 2021) was also performed in an antagonistic mode (Kayukawa et al., 2020). The library comprising 218 thousand items was exploited to yield 69 compounds structurally related to primary hits. These compounds inhibited JH I-dependent activation of the *k*JHRE-luciferase reporter, albeit in distinctive, competitive or noncompetitive modes. Interestingly, relatively minor structural differences turned some potential antagonists to activators of the *k*JHRE reporter. One of the competitive JH signalling inhibitors (JHSI48) was active at sub-micromolar concentrations in the cell-based assay and also reduced *Kr-h1* mRNA levels in isolated *Bombyx* tissues. Notably, JHSI48 induced precocious pupation in penultimate-instar larvae, a phenotype marking JH deficiency (Daimon et al., 2012, 2015; Tan et al., 2005). Whether JHSI48 is a genuine antagonist of the *Bombyx* JHR and whether it might be active in other insects should be addressed by future studies.

6. Conclusion

Together, the data reviewed in this chapter establish several key attributes defining the bHLH-PAS proteins Met/Gce as *bona fide* intracellular receptors for juvenile hormones and their synthetic mimics. The evidence is based at molecular, genetic, cellular, and organismal levels.

(1) JHs competitively bind to Met/Gce proteins from diverse species with nanomolar affinities; ligand binding is stereoselective and reliant on specific, conserved amino acid residues within the PAS-B domain of Met/Gce. Mutations of these residues block Met/Gce functions. (2) Met/Gce constitute JHR complexes that bind to JH-response regulatory DNA elements of direct target genes and induce their transcription in a manner dependent on JH binding to Met/Gce. (3) Met/Gce and its partner protein Tai are sufficient to confer JH responsiveness to mammalian cells. (4) Aberrant phenotypes caused by loss of Met/Gce function match specific roles of JH in insect development and reproduction; defects such as precocious metamorphosis phenocopy the deficiency of JH itself. (5) The capacity of Met/Gce to bind JH *in vivo* is required to sustain insect development.

While the discovery of the intracellular JHR has enabled considerable progress in diverse areas of arthropod research, much still remains to be learned about the molecular mechanism of JHR action. For instance, the exact mode of interaction and function of Tai, Cycle, Tgo, and other, hitherto unknown protein partners of Met/Gce within JHR complexes requires detailed mechanistic studies. The same applies to post-translational modification of the JHR, primarily phosphorylation induced by JH acting upon a hypothetical cell-membrane receptor. The role of phosphorylation in the nucleocytoplasmic transport of Met/Gce has not been addressed. Any ligand-induced conformational changes of Met/Gce leading to nuclear import and formation of an active JHR complex are completely unknown. At the moment, the nature of interactions of either native hormones or diverse synthetic JHR agonists with the ligand-binding pocket of JHR proteins is merely subject to computational modelling predictions. Clearly, the ultimate understanding of the architecture of the JHR complexes with their agonist ligands awaits structural resolution at the atomic level. Answering these questions will unleash the potential of the JHR as a target of novel compounds for selective and efficient intervention with insect development.

Acknowledgements

We thank Vlastimil Smykal for images in Fig. 3, Jean-Philippe Charles and Mirka Uhlirova for data in Figs. 4A and 5C, respectively, and Martina Hajduskova (www.bioGraphix.cz) for help with artwork. We are grateful to Ron Hill for permission to cite unpublished data obtained in collaboration with his former team at CSIRO in Australia. The authors' original work mentioned in this chapter was supported by the EXPRO project 20-05151X from the Czech Science Foundation.

References

Abdou, M.A., He, Q., Wen, D., Zyaan, O., Wang, J., Xu, J., Baumann, A.A., Joseph, J., Wilson, T.G., Li, S., Wang, J., 2011. *Drosophila* Met and Gce are partially redundant in transducing juvenile hormone action. Insect Biochem. Mol. Biol. 41, 938–945.

Ahmed, T.H., Saunders, T.R., Mullins, D., Rahman, M.Z., Zhu, J., 2020. Molecular action of pyriproxyfen: role of the Methoprene-tolerant protein in the pyriproxyfen-induced sterilization of adult female mosquitoes. PLoS Negl. Trop. Dis. 14, e0008669.

Arbeitman, M.N., Hogness, D.S., 2000. Molecular chaperones activate the *Drosophila* ecdysone receptor, an RXR heterodimer. Cell 101, 67–77.

Ashburner, M., Chihara, C., Meltzer, P., Richards, G., 1974. Temporal control of puffing activity in polytene chromosomes. Cold Spring Harb. Symp. Quant. Biol. 38, 655–662.

Ashok, M., Turner, C., Wilson, T.G., 1998. Insect juvenile hormone resistance gene homology with the bHLH-PAS family of transcriptional regulators. Proc. Natl. Acad. Sci. U. S. A. 95, 2761–2766.

Bai, J., Uehara, Y., Montell, D.J., 2000. Regulation of invasive cell behavior by taiman, a *Drosophila* protein related to AIB1, a steroid receptor coactivator amplified in breast cancer. Cell 103, 1047–1058.

Bajgar, A., Jindra, M., Dolezel, D., 2013. Autonomous regulation of the insect gut by circadian genes acting downstream of juvenile hormone signaling. Proc. Natl. Acad. Sci. U. S. A. 110, 4416–4421.

Baumann, A., Fujiwara, Y., Wilson, T.G., 2010. Evolutionary divergence of the paralogs Methoprene tolerant (Met) and germ cell expressed (gce) within the genus *Drosophila*. J. Insect Physiol. 56, 1445–1455.

Beischlag, T.V., Wang, S., Rose, D.W., Torchia, J., Reisz-Porszasz, S., Muhammad, K., Nelson, W.E., Probst, M.R., Rosenfeld, M.G., Hankinson, O., 2002. Recruitment of the NCoA/SRC-1/p160 family of transcriptional coactivators by the aryl hydrocarbon receptor/aryl hydrocarbon receptor nuclear translocator complex. Mol. Cell. Biol. 22, 4319–4333.

Beischlag, T.V., Morales, J.L., Hollingshead, B.D., Perdew, G.H., 2008. The aryl hydrocarbon receptor complex and the control of gene expression. Crit. Rev. Eukaryot. Gene Expr. 18, 207–250.

Bellés, X., 2020. Insect Metamorphosis: From Natural History to Regulation of Development and Evolution. Elsevier Inc., London.

Bender, M., Imam, F.B., Talbot, W.S., Ganetzky, B., Hogness, D.S., 1997. *Drosophila* ecdysone receptor mutations reveal functional differences among receptor isoforms. Cell 91, 777–788.

Berg, P., Pongratz, I., 2001. Differential usage of nuclear export sequences regulates intracellular localization of the dioxin (aryl hydrocarbon) receptor. J. Biol. Chem. 276, 43231–43238.

Bernardo, T.J., Dubrovsky, E.B., 2012. The *Drosophila* juvenile hormone receptor candidates methoprene-tolerant (MET) and germ cell-expressed (GCE) utilize a conserved LIXXL motif to bind the FTZ-F1 nuclear receptor. J. Biol. Chem. 287, 7821–7833.

Bilen, J., Atallah, J., Azanchi, R., Levine, J.D., Riddiford, L.M., 2013. Regulation of onset of female mating and sex pheromone production by juvenile hormone in *Drosophila melanogaster*. Proc. Natl. Acad. Sci. U. S. A. 110, 18321–18326.

Billas, I.M.L., Iwema, T., Garnier, J.-M., Mitschler, A., Rochel, N., Moras, D., 2003. Structural adaptability in the ligand-binding pocket of the ecdysone hormone receptor. Nature 426, 91–96.

Bittova, L., Jedlicka, P., Dracinsky, M., Kirubakaran, P., Vondrasek, J., Hanus, R., Jindra, M., 2019. Exquisite ligand stereoselectivity of a *Drosophila* juvenile hormone receptor contrasts with its broad agonist repertoire. J. Biol. Chem. 294, 410–423.

Bowers, W.S., Fales, H.M., Thompson, M.J., Uebel, E.C., 1966. Juvenile hormone: identification of an active compound from balsam fir. Science 154, 1020–1021.

Browning, C., McEwen, A.G., Mori, K., Yokoi, T., Moras, D., Nakagawa, Y., Billas, I.M.L., 2021. Nonsteroidal ecdysone receptor agonists use a water channel for binding to the ecdysone receptor complex EcR/USP. J. Pestic. Sci. 46, 88–100.

Butenandt, A., Karlson, P., 1954. Über die Isolierung eines Metamorphose-Hormons der Insecten in kristallisiertet Form. Z. Naturforsch. 9, 389–391.

Carmichael, J.A., Lawrence, M.C., Graham, L.D., Pilling, P.A., Epa, V.C., Noyce, L., Lovrecz, G., Winkler, D.A., Pawlak-Skrzecz, A., Eaton, R.E., Hannan, G.N., Hill, R.J., 2005. The X-ray structure of a hemipteran ecdysone receptor ligand-binding domain: comparison with a lepidopteran ecdysone receptor ligand-binding domain and implications for insecticide design. J. Biol. Chem. 280, 22258–22269.

Carney, G.E., Bender, M., 2000. The *Drosophila* ecdysone receptor (EcR) gene is required maternally for normal oogenesis. Genetics 154, 1203–1211.

Chang, M.-M., Wang, Y.-H., Yang, Q.-T., Wang, X.-L., Wang, M., Raikhel, A.S., Zou, Z., 2021. Regulation of antimicrobial peptides by juvenile hormone and its receptor, Methoprene-tolerant, in the mosquito *Aedes aegypti*. Insect Biochem. Mol. Biol. 128, 103509.

Charles, J.-P., Iwema, T., Epa, V.C., Takaki, K., Rynes, J., Jindra, M., 2011. Ligand-binding properties of a juvenile hormone receptor, Methoprene-tolerant. Proc. Natl. Acad. Sci. U. S. A. 108, 21128–21133.

Cherbas, P., Cherbas, L., Lee, S.-S., Nakanishi, K., 1988. 26-[^{125}I]Iodoponasterone A is a potent ecdysone and a sensitive radioligand for ecdysone receptors. Proc. Natl. Acad. Sci. U. S. A. 85, 2096–2100.

Cherbas, L., Lee, K., Cherbas, P., 1991. Identification of ecdysone response elements by analysis of the *Drosophila Eip28/29* gene. Genes Dev. 5, 120–131.

Corrada, D., Soshilov, A.A., Denison, M.S., Bonati, L., 2016. Deciphering dimerization modes of PAS domains: computational and experimental analyses of the AhR:ARNT complex reveal new insights into the mechanisms of AhR transformation. PLoS Comput. Biol. 12, e1004981.

Corrada, D., Denison, M.S., Bonati, L., 2017. Structural modeling of the AhR:ARNT complex in the bHLH–PASA–PASB region elucidates the key determinants of dimerization. Mol. Biosyst. 13, 981–990.
Cui, Y., Sui, Y., Xu, J., Zhu, F., Palli, S.R., 2014. Juvenile hormone regulates *Aedes aegypti* Krüppel homolog 1 through a conserved E box motif. Insect Biochem. Mol. Biol. 52, 23–32.
Daimon, T., Kozaki, T., Niwa, R., Kobayashi, I., Furuta, K., Namiki, T., Uchino, K., Banno, Y., Katsuma, S., Tamura, T., Mita, K., Sezutsu, H., Nakayama, M., Itoyama, K., Shimada, T., Shinoda, T., 2012. Precocious metamorphosis in the juvenile hormone-deficient mutant of the silkworm, *Bombyx mori*. PLoS Genet. 8, e1002486.
Daimon, T., Uchibori, M., Nakao, H., Sezutsu, H., Shinoda, T., 2015. Knockout silkworms reveal a dispensable role for juvenile hormones in holometabolous life cycle. Proc. Natl. Acad. Sci. U. S. A. 112, E4226–E4235.
Davey, K., 1996. Hormonal control of the follicular epithelium during vitellogenin uptake. Invertebr. Reprod. Dev. 30, 249–254.
Denison, M.S., 2010. Summary of Probe Development Efforts to Identify Activators of the Aryl Hydrocarbon Receptor (AHR). PubChem BioAssay Record for AID 602173, NCBI. https://pubchem.ncbi.nlm.nih.gov/bioassay/602173#section=Same-Project-BioAssays.
Denison, M.S., Faber, S.C., 2017. And now for something completely different: diversity in ligand-dependent activation of Ah receptor responses. Curr. Opin. Toxicol. 2, 124–131.
Denison, M.S., Nagy, S.R., 2003. Activation of the aryl hydrocarbon receptor by structurally diverse exogenous and endogenous ligands. Annu. Rev. Pharmacol. Toxicol. 43, 309–334.
Denison, M.S., Soshilov, A.A., He, G., DeGroot, D.E., Zhao, B., 2011. Exactly the same but different: promiscuity and diversity in the molecular mechanisms of action of the aryl hydrocarbon (dioxin) receptor. Toxicol. Sci. 124, 1–22.
Dolezel, D., 2015. Photoperiodic time measurement in insects. Curr. Opin. Insect Sci. 7, 98–103.
Dorn, S., Frischknecht, M.L., Martinez, V., Zurflüh, R., Fischer, U., 1981. A novel non-neurotoxic insecticide with a broad activity spectrum. Z. Pflanzenkrankh. Pflanzenschutz 88, 269–275.
Dubrovsky, E.B., Bernardo, T.J., 2014. The juvenile hormone receptor and molecular mechanisms of juvenile hormone action. Adv. Insect Physiol. 46, 305–388.
Erbel, P.J.A., Card, P.B., Karakuzu, O., Bruick, R.K., Gardner, K.H., 2003. Structural basis for PAS domain heterodimerization in the basic helix-loop-helix-PAS transcription factor hypoxia-inducible factor. Proc. Natl. Acad. Sci. U. S. A. 100, 15504–15509.
Evans, R.M., Mangelsdorf, D.J., 2014. Nuclear receptors, RXR, and the big bang. Cell 157, 255–266.
Feyereisen, R., Jindra, M., 2012. The silkworm coming of age-early. PLoS Genet. 8, e1002591.
Flatt, T., Tu, M.-P., Tatar, M., 2005. Hormonal pleiotropy and the juvenile hormone regulation of *Drosophila* development and life history. Bioessays 27, 999–1010.
Gijbels, M., Lenaerts, C., Vanden Broeck, J., Marchal, E., 2019. Juvenile hormone receptor Met is essential for ovarian maturation in the desert locust, *Schistocerca gregaria*. Sci. Rep. 9, 10797.
Gilbert, L.I., Granger, N.A., Roe, R.M., 2000. The juvenile hormones: historical facts and speculations on future research directions. Insect Biochem. Mol. Biol. 30, 617–644.
Godlewski, J., Wang, S., Wilson, T.G., 2006. Interaction of bHLH-PAS proteins involved in juvenile hormone reception in *Drosophila*. Biochem. Biophys. Res. Commun. 342, 1305–1311.
Goodman, W.G., Cusson, M., 2012. The juvenile hormones. In: Gilbert, L.I. (Ed.), Insect Endocrinology. Elsevier, Amsterdam, pp. 310–365.

Goodman, W., Schooley, D.A., Gilbert, L.I., 1978. Specificity of the juvenile hormone binding protein: the geometrical isomers of juvenile hormone I. Proc. Natl. Acad. Sci. U. S. A. 75, 185–189.

Graham, L.D., Johnson, W.M., Pawlak-Skrzecz, A., Eaton, R.E., Bliese, M., Howell, L., Hannan, G.N., Hill, R.J., 2007. Ligand binding by recombinant domains from insect ecdysone receptors. Insect Biochem. Mol. Biol. 37, 611–626.

Greb-Markiewicz, B., Kolonko, M., 2019. Subcellular localization signals of bHLH-PAS proteins: their significance, current state of knowledge and future perspectives. Int. J. Mol. Sci. 20, 4746.

Greb-Markiewicz, B., Orłowski, M., Dobrucki, J., Ożyhar, A., 2011. Sequences that direct subcellular traffic of the *Drosophila* methoprene-tolerant protein (MET) are located predominantly in the PAS domains. Mol. Cell. Endocrinol. 345, 16–26.

Greb-Markiewicz, B., Sadowska, D., Surgut, N., Godlewski, J., Zarębski, M., Ożyhar, A., 2015. Mapping of the sequences directing localization of the *Drosophila* germ cell-expressed protein (GCE). PLoS One 10, e0133307.

Gujar, H., Palli, S.R., 2016. Juvenile hormone regulation of female reproduction in the common bed bug, *Cimex lectularius*. Sci. Rep. 6, 35546.

Guo, W., Wu, Z., Song, J., Jiang, F., Wang, Z., Deng, S., Walker, V.K., Zhou, S., 2014. Juvenile hormone-receptor complex acts on mcm4 and mcm7 to promote polyploidy and vitellogenesis in the migratory locust. PLoS Genet. 10, e1004702.

Hall, B.L., Thummel, C.S., 1998. The RXR homolog ultraspiracle is an essential component of the *Drosophila* ecdysone receptor. Development 125, 4709–4717.

Hansen, I.A., Attardo, G.M., Rodriguez, S.D., Drake, L.L., 2014. Four-way regulation of mosquito yolk protein precursor genes by juvenile hormone-, ecdysone-, nutrient-, and insulin-like peptide signaling pathways. Front. Physiol. 5, 103.

Hatakoshi, M., Agui, N., Nakayama, I., 1986. 2-[1-methyl-2-(4-phenoxyphenoxy) ethoxy] pyridine as a new insect juvenile hormone analog: induction of supernumerary larvae in *Spodoptera litura* (Lepidoptera, Noctuidae). Appl. Entomol. Zool. 21, 351–353.

He, Q., Wen, D., Jia, Q., Cui, C., Wang, J., Palli, S.R., Li, S., 2014. Heat shock protein 83 (Hsp83) facilitates Methoprene-tolerant (Met) nuclear import to modulate juvenile hormone signaling. J. Biol. Chem. 289, 27874–27885.

He, Q., Zhang, Y., Zhang, X., Xu, D., Dong, W., Li, S., Wu, R., 2017. Nucleoporin Nup358 facilitates nuclear import of Methoprene-tolerant (Met) in an importin β- and Hsp83-dependent manner. Insect Biochem. Mol. Biol. 81, 10–18.

Helvig, C., Koener, J.F., Unnithan, G.C., Feyereisen, R., 2004. CYP15A1, the cytochrome P450 that catalyzes epoxidation of methyl farnesoate to juvenile hormone III in cockroach corpora allata. Proc. Natl. Acad. Sci. U. S. A. 101, 4024–4029.

Henrich, V.C., Szekely, A.A., Kim, S.J., Brown, N.E., Antoniewski, C., Hayden, M.A., Lepesant, J.A., Gilbert, L.I., 1994. Expression and function of the ultraspiracle (usp) gene during development of *Drosophila melanogaster*. Dev. Biol. 165, 38–52.

Henrick, C.A., 2007. Methoprene. J. Am. Mosq. Control Assoc. 23 (Suppl. 2), 225–239. AMCA bull. No. 7.

Henrick, C.A., Staal, G.B., Siddall, J.B., 1973. Alkyl 3,7,11-trimethyl-2,4-dodecadienoates, a new class of potent insect growth regulators with juvenile hormone activity. J. Agric. Food Chem. 21, 354–359.

Hill, R.J., Graham, L.D., Turner, K.A., Howell, L., Tohidi-Esfahani, D., Fernley, R., Grusovin, J., Ren, B., Pilling, P., Lu, L., Phan, T., Pollard, G.O.L., Pawlak-Skrzecz, A., Streltsov, V.A., Peat, T.S., Winkler, D.A., Lawrence, M.C., 2012. Structure and function of ecdysone receptors—interactions with ecdysteroids and synthetic agonists. Adv. Insect Physiol. 43, 229–351.

Hill, R.J., Billas, I.M.L., Bonneton, F., Graham, L.D., Lawrence, M.C., 2013. Ecdysone receptors: from the Ashburner model to structural biology. Annu. Rev. Entomol. 58, 251–271.

Hirano, M., Toyota, K., Ishibashi, H., Tominaga, N., Sato, T., Tatarazako, N., Iguchi, T., 2020. Molecular insights into structural and ligand binding features of Methoprene-tolerant in daphnids. Chem. Res. Toxicol. 33, 2785–2792.

Hiruma, K., Kaneko, Y., 2013. Hormonal regulation of insect metamorphosis with special reference to juvenile hormone biosynthesis. Curr. Top. Dev. Biol. 103, 73–100.

Ikuta, T., Eguchi, H., Tachibana, T., Yoneda, Y., Kawajiri, K., 1998. Nuclear localization and export signals of the human aryl hydrocarbon receptor. J. Biol. Chem. 273, 2895–2904.

Ikuta, T., Tachibana, T., Watanabe, J., Yoshida, M., Yoneda, Y., Kawajiri, K., 2000. Nucleocytoplasmic shuttling of the aryl hydrocarbon receptor. J. Biochem. 127, 503–509.

Ikuta, T., Kobayashi, Y., Kawajiri, K., 2004. Phosphorylation of nuclear localization signal inhibits the ligand-dependent nuclear import of aryl hydrocarbon receptor. Biochem. Biophys. Res. Commun. 317, 545–550.

Inui, T., Daimon, T., 2017. Implantation assays using the integument of early stage *Bombyx* larvae: insights into the mechanisms underlying the acquisition of competence for metamorphosis. J. Insect Physiol. 100, 35–42.

Ito-Harashima, S., Yagi, T., 2021. Reporter gene assays for screening and identification of novel molting hormone- and juvenile hormone-like chemicals. J. Pestic. Sci. 46, 29–42.

Iwema, T., Billas, I.M., Beck, Y., Bonneton, F., Nierengarten, H., Chaumot, A., Richards, G., Laudet, V., Moras, D., 2007. Structural and functional characterization of a novel type of ligand-independent RXR-USP receptor. EMBO J. 26, 3770–3782.

Jackson, D.P., Joshi, A.D., Elferink, C.J., 2015. Ah receptor pathway intricacies; signaling through diverse protein partners and DNA-motifs. Toxicol. Res. 4, 1143–1158.

Jindra, M., 2019. Where did the pupa come from? The timing of juvenile hormone signalling supports homology between stages of hemimetabolous and holometabolous insects. Philos. Trans. R. Soc. B 374, 20190064.

Jindra, M., Bittova, L., 2020. The juvenile hormone receptor as a target of juvenoid "insect growth regulators.". Arch. Insect Biochem. Physiol. 103, e21615.

Jindra, M., McKinstry, W., Nebl, T., Bittova, L., Ren, B., Xu, Z., Shaw, J., Phan, T., Lu, L., Low, J., Mackay, J., Song, X., Sparrow, L., Lovretz, G., Hill, R.J., 2021. Purification of an active juvenile hormone receptor complex and its post-translational modification. J. Biol. Chem. Submitted for publication.

Jindra, M., Palli, S.R., Riddiford, L.M., 2013. The juvenile hormone signaling pathway in insect development. Annu. Rev. Entomol. 58, 181–204.

Jindra, M., Bellés, X., Shinoda, T., 2015a. Molecular basis of juvenile hormone signaling. Curr. Opin. Insect Sci. 11, 39–46.

Jindra, M., Uhlirova, M., Charles, J.-P., Smykal, V., Hill, R.J., 2015b. Genetic evidence for function of the bHLH-PAS protein Gce/Met as a juvenile hormone receptor. PLoS Genet. 11, e1005394.

Jing, Y.-P., An, H., Zhang, S., Wang, N., Zhou, S., 2018. Protein kinase C mediates juvenile hormone-dependent phosphorylation of Na^+/K^+-ATPase to induce ovarian follicular patency for yolk protein uptake. J. Biol. Chem. 293, 20112–20122.

Kakaley, E.K.M., Wang, H.Y., LeBlanc, G.A., 2017. Agonist-mediated assembly of the crustacean methyl farnesoate receptor. Sci. Rep. 7, 45071.

Kayukawa, T., Shinoda, T., 2015. Functional characterization of two paralogous JH receptors, methoprene-tolerant 1 and 2, in the silkworm, *Bombyx mori* (Lepidoptera: Bombycidae). Appl. Entomol. Zool. 50, 383–391.

Kayukawa, T., Minakuchi, C., Namiki, T., Togawa, T., Yoshiyama, M., Kamimura, M., Mita, K., Imanishi, S., Kiuchi, M., Ishikawa, Y., Shinoda, T., 2012. Transcriptional regulation of juvenile hormone-mediated induction of Krüppel homolog 1, a repressor of insect metamorphosis. Proc. Natl. Acad. Sci. U. S. A. 109, 11729–11734.

Kayukawa, T., Tateishi, K., Shinoda, T., 2013. Establishment of a versatile cell line for juvenile hormone signaling analysis in *Tribolium castaneum*. Sci. Rep. 3, 1570.

Kayukawa, T., Murata, M., Kobayashi, I., Muramatsu, D., Okada, C., Uchino, K., Sezutsu, H., Kiuchi, M., Tamura, T., Hiruma, K., Ishikawa, Y., Shinoda, T., 2014. Hormonal regulation and developmental role of Krüppel homolog 1, a repressor of metamorphosis, in the silkworm *Bombyx mori*. Dev. Biol. 388, 48–56.

Kayukawa, T., Jouraku, A., Ito, Y., Shinoda, T., 2017. Molecular mechanism underlying juvenile hormone-mediated repression of precocious larval-adult metamorphosis. Proc. Natl. Acad. Sci. U. S. A. 114, 1057–1062.

Kayukawa, T., Furuta, K., Nagamine, K., Shinoda, T., Yonesu, K., Okabe, T., 2020. Identification of a juvenile-hormone signaling inhibitor via high-throughput screening of a chemical library. Sci. Rep. 10, 18413.

Kayukawa, T., Furuta, K., Yonesu, K., Okabe, T., 2021. Identification of novel juvenile-hormone signaling activators *via* high-throughput screening with a chemical library. J. Pestic. Sci. 46, 53–59.

Kazlauskas, A., Sundström, S., Poellinger, L., Pongratz, I., 2001. The hsp90 chaperone complex regulates intracellular localization of the dioxin receptor. Mol. Cell. Biol. 21, 2594–2607.

Kewley, R.J., Whitelaw, M.L., Chapman-Smith, A., 2004. The mammalian basic helix-loop-helix/PAS family of transcriptional regulators. Int. J. Biochem. Cell Biol. 36, 189–204.

King-Jones, K., Thummel, C.S., 2005. Nuclear receptors—a perspective from *Drosophila*. Nat. Rev. Genet. 6, 311–323.

Koelle, M.R., Talbot, W.S., Segraves, W.A., Bender, M.T., Cherbas, P., Hogness, D.S., 1991. The *Drosophila* EcR gene encodes an ecdysone receptor, a new member of the steroid receptor superfamily. Cell 67, 59–77.

Kolonko, M., Ożga, K., Hołubowicz, R., Taube, M., Kozak, M., Ożyhar, A., Greb-Markiewicz, B., 2016. Intrinsic disorder of the C-terminal domain of *Drosophila* Methoprene-tolerant protein. PLoS One 11, e0162950.

Kolonko, M., Bystranowska, D., Taube, M., Kozak, M., Bostock, M., Popowicz, G., Ożyhar, A., Greb-Markiewicz, B., 2020. The intrinsically disordered region of GCE protein adopts a more fixed structure by interacting with the LBD of the nuclear receptor FTZ-F1. Cell Commun. Signal 18, 180.

Konopova, B., Jindra, M., 2007. Juvenile hormone resistance gene Methoprene-tolerant controls entry into metamorphosis in the beetle *Tribolium castaneum*. Proc. Natl. Acad. Sci. U. S. A. 104, 10488–10493.

Konopova, B., Jindra, M., 2008. Broad-complex acts downstream of Met in juvenile hormone signaling to coordinate primitive holometabolan metamorphosis. Development 135, 559–568.

Konopova, B., Smykal, V., Jindra, M., 2011. Common and distinct roles of juvenile hormone signaling genes in metamorphosis of holometabolous and hemimetabolous insects. PLoS One 6, e28728.

Kumar, M.B., Perdew, G.H., 1999. Nuclear receptor coactivator SRC-1 interacts with the Q-rich subdomain of the AhR and modulates its transactivation potential. Gene Expr. 8, 273–286.

Lee, S.-H., Oh, H.-W., Fang, Y., An, S.-B., Park, D.-S., Song, H.-H., Oh, S.-R., Kim, S.-Y., Kim, S., Kim, N., Raikhel, A.S., Je, Y.H., Shin, S.W., 2015. Identification of plant compounds that disrupt the insect juvenile hormone receptor complex. Proc. Natl. Acad. Sci. U. S. A. 112, 1733–1738.

Lee, S.-H., Ha, K.B., Park, D.H., Fang, Y., Kim, J.H., Park, M.G., Woo, R.M., Kim, W.J., Park, I.-K., Choi, J.Y., Je, Y.H., 2018. Plant-derived compounds regulate formation of the insect juvenile hormone receptor complex. Pestic. Biochem. Physiol. 150, 27–32.

Lees, M.J., Whitelaw, M.L., 1999. Multiple roles of ligand in transforming the dioxin receptor to an active basic helix-loop-helix/PAS transcription factor complex with the nuclear protein Arnt. Mol. Cell. Biol. 19, 5811–5822.

Li, M., Mead, E.A., Zhu, J., 2011. Heterodimer of two bHLH-PAS proteins mediates juvenile hormone-induced gene expression. Proc. Natl. Acad. Sci. U. S. A. 108, 638–643.

Li, M., Liu, P., Wiley, J.D., Ojani, R., Bevan, D.R., Li, J., Zhu, J., 2014. A steroid receptor coactivator acts as the DNA-binding partner of the methoprene-tolerant protein in regulating juvenile hormone response genes. Mol. Cell. Endocrinol. 394, 47–58.

Li, X., Chen, T., Han, Y., Huang, M., Jiang, H., Huang, J., Tao, M., Xu, R., Xie, Q., Su, S., 2021a. Potential role of Methoprene-tolerant (Met) in methyl farnesoate-mediated vitellogenesis in the Chinese mitten crab (*Eriocheir sinensis*). Comp. Biochem. Physiol. B Biochem. Mol. Biol. 252, 110524.

Li, Y.-X., Wang, D., Zhao, W.-L., Zhang, J.-Y., Kang, X.-L., Li, Y.-L., Zhao, X.-F., 2021b. Juvenile hormone induces methoprene-tolerant 1 phosphorylation to increase interaction with Taiman in *Helicoverpa armigera*. Insect Biochem. Mol. Biol. 130, 103519.

Liszekova, D., Polakovicova, M., Beno, M., Farkas, R., 2009. Molecular determinants of juvenile hormone action as revealed by 3D QSAR analysis in *Drosophila*. PLoS One 4, e6001.

Liu, Y., Sheng, Z., Liu, H., Wen, D., He, Q., Wang, S., Shao, W., Jiang, R.J., An, S., Sun, Y., Bendena, W.G., Wang, J., Gilbert, L.I., Wilson, T.G., Song, Q., Li, S., 2009. Juvenile hormone counteracts the bHLH-PAS transcription factors MET and GCE to prevent caspase-dependent programmed cell death in *Drosophila*. Development 136, 2015–2025.

Liu, W., Zhang, F.-X., Cai, M.-J., Zhao, W.-L., Li, X.-R., Wang, J.-X., Zhao, X.-F., 2013. The hormone-dependent function of Hsp90 in the crosstalk between 20-hydroxyecdysone and juvenile hormone signaling pathways in insects is determined by differential phosphorylation and protein interactions. Biochim. Biophys. Acta 1830, 5184–5192.

Liu, P., Peng, H.-J., Zhu, J., 2015. Juvenile hormone-activated phospholipase C pathway enhances transcriptional activation by the methoprene-tolerant protein. Proc. Natl. Acad. Sci. U. S. A. 112, E1871–E1879.

Liu, W., Li, Y., Zhu, L., Zhu, F., Lei, C.-L., Wang, X.-P., 2016. Juvenile hormone facilitates the antagonism between adult reproduction and diapause through the methoprene-tolerant gene in the female *Colaphellus bowringi*. Insect Biochem. Mol. Biol. 74, 50–60.

Liu, P., Fu, X., Zhu, J., 2018. Juvenile hormone-regulated alternative splicing of the taiman gene primes the ecdysteroid response in adult mosquitoes. Proc. Natl. Acad. Sci. U. S. A. 115, E7738–E7747.

Lozano, J., Bellés, X., 2011. Conserved repressive function of Krüppel homolog 1 on insect metamorphosis in hemimetabolous and holometabolous species. Sci. Rep. 1, 163.

Lozano, J., Bellés, X., 2014. Role of methoprene-tolerant (Met) in adult morphogenesis and in adult ecdysis of *Blattella germanica*. PLoS One 9, e103614.

Lozano, J., Kayukawa, T., Shinoda, T., Bellés, X., 2014. A role for taiman in insect metamorphosis. PLoS Genet. 10, e1004769.

Madhavan, K., 1973. Morhogenetic effects of juvenile hormone and juvenile hormone mimics on adult development of *Drosophila*. J. Insect Physiol. 19, 441–453.

Mangelsdorf, D.J., Thummel, C., Beato, M., Herrlich, P., Schütz, G., Umesono, K., Blumberg, B., Kastner, P., Mark, M., Chambon, P., Evans, R.M., 1995. The nuclear receptor superfamily: the second decade. Cell 83, 835–839.

Marchal, E., Hult, E.F., Huang, J., Pang, Z., Stay, B., Tobe, S.S., 2014. Methoprene-tolerant (Met) knockdown in the adult female cockroach, *Diploptera punctata* completely inhibits ovarian development. PLoS One 9, e106737.

Masuoka, Y., Yaguchi, H., Suzuki, R., Maekawa, K., 2015. Knockdown of the juvenile hormone receptor gene inhibits soldier-specific morphogenesis in the damp-wood termite *Zootermopsis nevadensis* (Isoptera: Archotermopsidae). Insect Biochem. Mol. Biol. 64, 25–31.

Masuoka, Y., Toga, K., Nalepa, C.A., Maekawa, K., 2018. A crucial caste regulation gene detected by comparing termites and sister group cockroaches. Genetics 209, 1225–1234.

Minakuchi, C., Riddiford, L., 2006. Insect juvenile hormone action as a potential target of pest management. J. Pestic. Sci. 31, 77–84.

Minakuchi, C., Namiki, T., Yoshiyama, M., Shinoda, T., 2008a. RNAi-mediated knockdown of juvenile hormone acid O-methyltransferase gene causes precocious metamorphosis in the red flour beetle *Tribolium castaneum*. FEBS J. 275, 2919–2931.

Minakuchi, C., Zhou, X., Riddiford, L.M., 2008b. Krüppel homolog 1 (Kr-h1) mediates juvenile hormone action during metamorphosis of *Drosophila melanogaster*. Mech. Dev. 125, 91–105.

Minakuchi, C., Namiki, T., Shinoda, T., 2009. Krüppel homolog 1, an early juvenile hormone-response gene downstream of Methoprene-tolerant, mediates its anti-metamorphic action in the red flour beetle *Tribolium castaneum*. Dev. Biol. 325, 341–350.

Minakuchi, C., Ishii, F., Washidu, Y., Ichikawa, A., Tanaka, T., Miura, K., Shinoda, T., 2015. Expressional and functional analysis of CYP15A1, a juvenile hormone epoxidase, in the red flour beetle *Tribolium castaneum*. J. Insect Physiol. 80, 61–70.

Miura, K., Oda, M., Makita, S., Chinzei, Y., 2005. Characterization of the *Drosophila* Methoprene -tolerant gene product. FEBS J. 272, 1169–1178.

Miyakawa, H., Iguchi, T., 2017. Comparative luciferase assay for establishing reliable in vitro screening system of juvenile hormone agonists. J. Appl. Toxicol. 37, 1082–1090.

Miyakawa, H., Toyota, K., Hirakawa, I., Ogino, Y., Miyagawa, S., Oda, S., Tatarazako, N., Miura, T., Colbourne, J.K., Iguchi, T., 2013. A mutation in the receptor Methoprene-tolerant alters juvenile hormone response in insects and crustaceans. Nat. Commun. 4, 1856.

Miyakawa, H., Sato, T., Song, Y., Tollefsen, K.E., Iguchi, T., 2018. Ecdysteroid and juvenile hormone biosynthesis, receptors and their signaling in the freshwater microcrustacean *Daphnia*. J. Steroid Biochem. Mol. Biol. 184, 62–68.

Motto, I., Bordogna, A., Soshilov, A.A., Denison, M.S., Bonati, L., 2011. New aryl hydrocarbon receptor homology model targeted to improve docking reliability. J. Chem. Inf. Model. 51, 2868–2881.

Nijhout, H.F., Riddiford, L.M., Mirth, C., Shingleton, A.W., Suzuki, Y., Callier, V., 2014. The developmental control of size in insects. Wiley Interdiscip. Rev. Dev. Biol. 3, 113–134.

Noriega, F.G., Shah, D.K., Wells, M.A., 1997. Juvenile hormone controls early trypsin gene transcription in the midgut of *Aedes aegypti*. Insect Mol. Biol. 6, 63–66.

Oh, H.-W., Yun, C.-S., Jeon, J.H., Kim, J.-A., Park, D.-S., Ryu, H.W., Oh, S.-R., Song, H.-H., Shin, Y., Jung, C.S., Shin, S.W., 2017. Conifer diterpene resin acids disrupt juvenile hormone-mediated endocrine regulation in the Indian meal moth *Plodia interpunctella*. J. Chem. Ecol. 43, 703–711.

Ohtake, F., Takeyama, K.-I., Matsumoto, T., Kitagawa, H., Yamamoto, Y., Nohara, K., Tohyama, C., Krust, A., Mimura, J., Chambon, P., Yanagisawa, J., Fujii-Kuriyama, Y., Kato, S., 2003. Modulation of oestrogen receptor signalling by association with the activated dioxin receptor. Nature 423, 545–550.

Ohtake, F., Baba, A., Takada, I., Okada, M., Iwasaki, K., Miki, H., Takahashi, S., Kouzmenko, A., Nohara, K., Chiba, T., Fujii-Kuriyama, Y., Kato, S., 2007. Dioxin receptor is a ligand-dependent E3 ubiquitin ligase. Nature 446, 562–566.

Ojani, R., Liu, P., Fu, X., Zhu, J., 2016. Protein kinase C modulates transcriptional activation by the juvenile hormone receptor methoprene-tolerant. Insect Biochem. Mol. Biol. 70, 44–52.

Oro, A.E., McKeown, M., Evans, R.M., 1992. The *Drosophila* retinoid X receptor homolog ultraspiracle functions in both female reproduction and eye morphogenesis. Development 115, 449–462.

Pandini, A., Denison, M.S., Song, Y., Soshilov, A.A., Bonati, L., 2007. Structural and functional characterization of the aryl hydrocarbon receptor ligand binding domain by homology modeling and mutational analysis. Biochemistry 46, 696–708.

Pandini, A., Soshilov, A.A., Song, Y., Zhao, J., Bonati, L., Denison, M.S., 2009. Detection of the TCDD binding-fingerprint within the Ah receptor ligand binding domain by structurally driven mutagenesis and functional analysis. Biochemistry 48, 5972–5983.

Parthasarathy, R., Farkas, R., Palli, S.R., 2012. Recent progress in juvenile hormone analogs (JHA) research. Adv. Insect Physiol. 43, 353–436.

Pener, M.P., Dhadialla, T.S., 2012. An overview of insect growth disruptors; applied aspects. Adv. Insect Physiol. 43, 1–162.

Picard, D., 2006. Chaperoning steroid hormone action. Trends Endocrinol. Metab. 17, 229–235.

Pongratz, I., Mason, G.G., Poellinger, L., 1992. Dual roles of the 90-kDa heat shock protein hsp90 in modulating functional activities of the dioxin receptor. J. Biol. Chem. 267, 13728–13734.

Pursley, S., Ashok, M., Wilson, T.G., 2000. Intracellular localization and tissue specificity of the Methoprene-tolerant (Met) gene product in *Drosophila melanogaster*. Insect Biochem. Mol. Biol. 30, 839–845.

Qu, Z., Kenny, N.J., Lam, H.M., Chan, T.F., Chu, K.H., Bendena, W.G., Tobe, S.S., Hui, J.H.L., 2015. How did arthropod sesquiterpenoids and ecdysteroids arise? Comparison of hormonal pathway genes in noninsect arthropod genomes. Genome Biol. Evol. 7, 1951–1959.

Rahman, M.M., Franch-Marro, X., Maestro, J.L., Martin, D., Casali, A., 2017. Local juvenile hormone activity regulates gut homeostasis and tumor growth in adult *Drosophila*. Sci. Rep. 7, 11677.

Reiff, T., Jacobson, J., Cognigni, P., Antonello, Z., Ballesta, E., Tan, K.J., Yew, J.Y., Dominguez, M., Miguel-Aliaga, I., 2015. Endocrine remodelling of the adult intestine sustains reproduction in *Drosophila*. Elife 4, 06930.

Ren, B., Peat, T.S., Streltsov, V.A., Pollard, M., Fernley, R., Grusovin, J., Seabrook, S., Pilling, P., Phan, T., Lu, L., Lovrecz, G.O., Graham, L.D., Hill, R.J., 2014. Unprecedented conformational flexibility revealed in the ligand-binding domains of the *Bovicola ovis* ecdysone receptor (EcR) and ultraspiracle (USP) subunits. Acta Crystallogr. D Biol. Crystallogr. 70, 1954–1964.

Restifo, L.L., Wilson, T.G., 1998. A juvenile hormone agonist reveals distinct developmental pathways mediated by ecdysone-inducible broad complex transcription factors. Dev. Genet. 22, 141–159.

Richard, D.S., Applebaum, S.W., Sliter, T.J., Baker, F.C., Schooley, D.A., Reuter, C.C., Henrich, V.C., Gilbert, L.I., 1989. Juvenile hormone bisepoxide biosynthesis in vitro by the ring gland of *Drosophila melanogaster*: a putative juvenile hormone in the higher Diptera. Proc. Natl. Acad. Sci. U. S. A. 86, 1421–1425.

Riddiford, L., 1994. Cellular and molecular actions of juvenile hormone. I. General considerations and premetamorphic actions. Adv. Insect Physiol. 24, 213–274.

Riddiford, L.M., 2020. *Rhodnius*, golden oil, and Met: a history of juvenile hormone research. Front. Cell Dev. Biol. 8, 679.

Riddiford, L.M., Ashburner, M., 1991. Effects of juvenile hormone mimics on larval development and metamorphosis of *Drosophila melanogaster*. Gen. Comp. Endocrinol. 82, 172–183.

Riddiford, L.M., Truman, J.W., Mirth, C.K., Shen, Y.C., 2010. A role for juvenile hormone in the prepupal development of *Drosophila melanogaster*. Development 137, 1117–1126.

Röller, H., Dahm, K.H., Sweeley, C.C., Trost, B.M., 1967. Die Struktur des Juvenilhormones. Angew. Chem. 4, 190–191.

Roy, S., Saha, T.T., Zou, Z., Raikhel, A.S., 2018. Regulatory pathways controlling female insect reproduction. Annu. Rev. Entomol. 63, 489–511.

Saha, T.T., Shin, S.W., Dou, W., Roy, S., Zhao, B., Hou, Y., Wang, X.-L., Zou, Z., Girke, T., Raikhel, A.S., 2016. Hairy and Groucho mediate the action of juvenile hormone receptor Methoprene-tolerant in gene repression. Proc. Natl. Acad. Sci. U. S. A. 113, E735–E743.

Santos, C.G., Humann, F.C., Hartfelder, K., 2019. Juvenile hormone signaling in insect oogenesis. Curr. Opin. Insect Sci. 31, 43–48.

Schenk, S., Krauditsch, C., Frühauf, P., Gerner, C., Raible, F., 2016. Discovery of methylfarnesoate as the annelid brain hormone reveals an ancient role of sesquiterpenoids in reproduction. Elife 5, e17126.

Scheuermann, T.H., Tomchick, D.R., Machius, M., Guo, Y., Bruick, R.K., Gardner, K.H., 2009. Artificial ligand binding within the HIF2alpha PAS-B domain of the HIF2 transcription factor. Proc. Natl. Acad. Sci. U. S. A. 106, 450–455.

Schulte, K.W., Green, E., Wilz, A., Platten, M., Daumke, O., 2017. Structural basis for aryl hydrocarbon receptor-mediated gene activation. Structure 25, 1025–1033.e3.

Schwenke, R.A., Lazzaro, B.P., 2017. Juvenile hormone suppresses resistance to infection in mated female *Drosophila melanogaster*. Curr. Biol. 27, 596–601.

Seok, S.-H., Lee, W., Jiang, L., Molugu, K., Zheng, A., Li, Y., Park, S., Bradfield, C.A., Xing, Y., 2017. Structural hierarchy controlling dimerization and target DNA recognition in the AHR transcriptional complex. Proc. Natl. Acad. Sci. U. S. A. 114, 5431–5436.

Sevala, V.L., Davey, K., 1989. Action of juvenile hormone on the follicle cells of *Rhodnius prolixus*: evidence for a novel regulatory mechanism involving protein kinase C. Experientia 45, 355–356.

Sheng, Z., Xu, J., Bai, H., Zhu, F., Palli, S.R., 2011. Juvenile hormone regulates vitellogenin gene expression through insulin-like peptide signaling pathway in the red flour beetle, *Tribolium castaneum*. J. Biol. Chem. 286, 41924–41936.

Shin, S.W., Zou, Z., Saha, T.T., Raikhel, A.S., 2012. bHLH-PAS heterodimer of methoprene-tolerant and Cycle mediates circadian expression of juvenile hormone-induced mosquito genes. Proc. Natl. Acad. Sci. U. S. A. 109, 16576–16581.

Shin, S.W., Jeon, J.H., Jeong, S.A., Kim, J.-A., Park, D.-S., Shin, Y., Oh, H.-W., 2018. A plant diterpene counteracts juvenile hormone-mediated gene regulation during *Drosophila melanogaster* larval development. PLoS One 13, e0200706.

Shinoda, T., Itoyama, K., 2003. Juvenile hormone acid methyltransferase: a key regulatory enzyme for insect metamorphosis. Proc. Natl. Acad. Sci. U. S. A. 100, 11986–11991.

Sláma, K., Williams, C.M., 1965. Juvenile hormone activity for the bug *Pyrrhocoris apterus*. Proc. Natl. Acad. Sci. U. S. A. 54, 411–414.

Sláma, K., Romanuk, M., Sorm, F., 1974. Insect Hormones and Bioanalogues. Springer Verlag, New York.

Smykal, V., Raikhel, A.S., 2015. Nutritional control of insect reproduction. Curr. Opin. Insect Sci. 11, 31–38.

Smykal, V., Bajgar, A., Provaznik, J., Fexova, S., Buricova, M., Takaki, K., Hodkova, M., Jindra, M., Dolezel, D., 2014a. Juvenile hormone signaling during reproduction and development of the linden bug, *Pyrrhocoris apterus*. Insect Biochem. Mol. Biol. 45, 69–76.

Smykal, V., Daimon, T., Kayukawa, T., Takaki, K., Shinoda, T., Jindra, M., 2014b. Importance of juvenile hormone signaling arises with competence of insect larvae to metamorphose. Dev. Biol. 390, 221–230.
Soshilov, A., Denison, M.S., 2008. Role of the Per/Arnt/Sim domains in ligand-dependent transformation of the aryl hydrocarbon receptor. J. Biol. Chem. 283, 32995–33005.
Soshilov, A., Denison, M.S., 2011. Ligand displaces heat shock protein 90 from overlapping binding sites within the aryl hydrocarbon receptor ligand-binding domain. J. Biol. Chem. 286, 35275–35282.
Soshilov, A.A., Denison, M.S., 2014. Ligand promiscuity of aryl hydrocarbon receptor agonists and antagonists revealed by site-directed mutagenesis. Mol. Cell. Biol. 34, 1707–1719.
Soshilov, A.A., Motta, S., Bonati, L., Denison, M.S., 2020. Transitional states in ligand-dependent transformation of the aryl hydrocarbon receptor into its DNA-binding form. Int. J. Mol. Sci. 21, 2474.
Stejskalova, L., Dvorak, Z., Pavek, P., 2011. Endogenous and exogenous ligands of aryl hydrocarbon receptor: current state of art. Curr. Drug Metab. 12, 198–212.
Stockinger, B., Di Meglio, P., Gialitakis, M., Duarte, J.H., 2014. The aryl hydrocarbon receptor: multitasking in the immune system. Annu. Rev. Immunol. 32, 403–432.
Swevers, L., 2019. An update on ecdysone signaling during insect oogenesis. Curr. Opin. Insect Sci. 31, 8–13.
Tagliabue, S.G., Faber, S.C., Motta, S., Denison, M.S., Bonati, L., 2019. Modeling the binding of diverse ligands within the Ah receptor ligand binding domain. Sci. Rep. 9, 10693.
Tan, A., Tanaka, H., Tamura, T., Shiotsuki, T., 2005. Precocious metamorphosis in transgenic silkworms overexpressing juvenile hormone esterase. Proc. Natl. Acad. Sci. U. S. A. 102, 11751–11756.
Thomas, H.E., Stunnenberg, H.G., Stewart, A.F., 1993. Heterodimerization of the *Drosophila* ecdysone receptor with retinoid X receptor and ultraspiracle. Nature 362, 471–475.
Thummel, C.S., 1996. Flies on steroids—*Drosophila* metamorphosis and the mechanisms of steroid hormone action. Trends Genet. 12, 306–310.
Tkachenko, A., Henkler, F., Brinkmann, J., Sowada, J., Genkinger, D., Kern, C., Tralau, T., Luch, A., 2016. The Q-rich/PST domain of the AHR regulates both ligand-induced nuclear transport and nucleocytoplasmic shuttling. Sci. Rep. 6, 32009.
Tomoyasu, Y., Denell, R.E., 2004. Larval RNAi in *Tribolium* (Coleoptera) for analyzing adult development. Dev. Genes Evol. 214, 575–578.
Touhara, K., Lerro, K.A., Bonning, B.C., Hammock, B.D., Prestwich, G.D., 1993. Ligand binding by a recombinant insect juvenile hormone binding protein. Biochemistry 32, 2068–2075.
Truman, J.W., 2019. The evolution of insect metamorphosis. Curr. Biol. 29, R1252–R1268.
Tsang, S.S.K., Law, S.T.S., Li, C., Qu, Z., Bendena, W.G., Tobe, S.S., Hui, J.H.L., 2020. Diversity of insect sesquiterpenoid regulation. Front. Genet. 11, 1027.
Ureña, E., Chafino, S., Manjón, C., Franch-Marro, X., Martín, D., 2016. The occurrence of the holometabolous pupal stage requires the interaction between E93, Krüppel-homolog 1 and Broad-complex. PLoS Genet. 12, e1006020.
Villalobos-Sambucaro, M.A.J., Riccillo, F.L., Calderón-Fernández, G.M., Sterkel, M., Diambra, L.A., Ronderos, J.R., 2015. Genomic and functional characterization of a methoprene-tolerant gene in the kissing-bug *Rhodnius prolixus*. Gen. Comp. Endocrinol. 216, 1–8.
Wang, X., Hou, Y., Saha, T.T., Pei, G., Raikhel, A.S., Zou, Z., 2017a. Hormone and receptor interplay in the regulation of mosquito lipid metabolism. Proc. Natl. Acad. Sci. U. S. A. 114, E2709–E2718.

Wang, Z., Yang, L., Song, J., Kang, L., Zhou, S., 2017b. An isoform of Taiman that contains a PRD-repeat motif is indispensable for transducing the vitellogenic juvenile hormone signal in *Locusta migratoria*. Insect Biochem. Mol. Biol. 82, 31–40.

Wen, D., Rivera-Perez, C., Abdou, M., Jia, Q., He, Q., Liu, X., Zyaan, O., Xu, J., Bendena, W.G., Tobe, S.S., Noriega, F.G., Palli, S.R., Wang, J., Li, S., 2015. Methyl farnesoate plays a dual role in regulating *Drosophila* metamorphosis. PLoS Genet. 11, e1005038.

Whitelaw, M.L., McGuire, J., Picard, D., Gustafsson, J.A., Poellinger, L., 1995. Heat shock protein hsp90 regulates dioxin receptor function in vivo. Proc. Natl. Acad. Sci. U. S. A. 92, 4437–4441.

Wigglesworth, V., 1934. The physiology of ecdysis in *Rhodnius prolixus* (Hemiptera). II. Factors controlling moulting and "metamorphosis.". Q. J. Microsc. Sci. 77, 191–222.

Wigglesworth, V., 1936. The function of the corpus allatum in the growth and reproduction of *Rhodnius prolixus* (Hemiptera). Q. J. Microsc. Sci. 79, 91–121.

Wigglesworth, V., 1940. The determination of characters at metamorphosis of *Rhodnius prolixus* (Hemiptera). J. Exp. Biol. 17, 201–223.

Wigglesworth, V., 1958. Some methods for assaying extracts of the juvenile hormone in insects. J. Insect Physiol. 2, 73–84.

Wigglesworth, V., 1969. Chemical structure and juvenile hormone activity: comparative tests on *Rhodnius prolixus*. J. Insect Physiol. 15, 73–94.

Williams, C.M., 1956. The juvenile hormone of insects. Nature 178, 212–213.

Williams, C.M., 1967. Third-generation pesticides. Sci. Am. 217, 13–17.

Wilson, T.G., Ashok, M., 1998. Insecticide resistance resulting from an absence of target-site gene product. Proc. Natl. Acad. Sci. U. S. A. 95, 14040–14044.

Wilson, T.G., Fabian, J., 1986. A *Drosophila melanogaster* mutant resistant to a chemical analog of juvenile hormone. Dev. Biol. 118, 190–201.

Wilson, T.G., Wang, S., Beno, M., Farkas, R., 2006. Wide mutational spectrum of a gene involved in hormone action and insecticide resistance in *Drosophila melanogaster*. Mol. Genet. Genomics 276, 294–303.

Wu, Z., Yang, L., He, Q., Zhou, S., 2021. Regulatory mechanisms of vitellogenesis in insects. Front. Cell Dev. Biol. 8, 593613.

Wyatt, G., Davey, K., 1996. Cellular and molecular actions of juvenile hormone. II. Roles of juvenile hormone in adult insects. Adv. Insect Physiol. 26, 1–155.

Xu, J., Sheng, Z., Palli, S.R., 2013. Juvenile hormone and insulin regulate trehalose homeostasis in the red flour beetle, *Tribolium castaneum*. PLoS Genet. 9, e1003535.

Yamamoto, K., Chadarevian, A., Pellegrini, M., 1988. Juvenile hormone action mediated in male accessory glands of *Drosophila* by calcium and kinase C. Science 239, 916–919.

Yamanaka, N., Rewitz, K.F., O'Connor, M.B., 2013. Ecdysone control of developmental transitions: lessons from *Drosophila* research. Annu. Rev. Entomol. 58, 497–516.

Yao, T.P., Forman, B.M., Jiang, Z., Cherbas, L., Chen, J.D., McKeown, M., Cherbas, P., Evans, R.M., 1993. Functional ecdysone receptor is the product of EcR and ultraspiracle genes. Nature 366, 476–479.

Yokoi, T., Nabe, T., Ishizuka, C., Hayashi, K., Ito-Harashima, S., Yagi, T., Nakagawa, Y., Miyagawa, H., 2020. Transcription-inducing activity of natural and synthetic juvenile hormone agonists through the *Drosophila* Methoprene-tolerant protein. Pest Manag. Sci. 76, 2316–2323.

Yokoi, T., Nabe, T., Horoiwa, S., Hayashi, K., Ito-Harashima, S., Yagi, T., Nakagawa, Y., Miyagawa, H., 2021. Virtual screening identifies a novel piperazine-based insect juvenile hormone agonist. J. Pestic. Sci. 46, 68–74.

Zaoral, M., Slama, K., 1970. Peptides with juvenile hormone activity. Science 170, 92–93.

Zhang, S., Rowlands, C., Safe, S., 2008. Ligand-dependent interactions of the Ah receptor with coactivators in a mammalian two-hybrid assay. Toxicol. Appl. Pharmacol. 227, 196–206.

Zhang, Z., Xu, J., Sheng, Z., Sui, Y., Palli, S.R., 2011. Steroid receptor co-activator is required for juvenile hormone signal transduction through a bHLH-PAS transcription factor, methoprene tolerant. J. Biol. Chem. 286, 8437–8447.

Zhou, X., Riddiford, L.M., 2002. Broad specifies pupal development and mediates the "status quo" action of juvenile hormone on the pupal-adult transformation in *Drosophila* and *Manduca*. Development 129, 2259–2269.

Zhu, J., Chen, L., Sun, G., Raikhel, A.S., 2006. The competence factor beta Ftz-F1 potentiates ecdysone receptor activity via recruiting a p160/SRC coactivator. Mol. Cell. Biol. 26, 9402–9412.

Zhu, J., Busche, J.M., Zhang, X., 2010. Identification of juvenile hormone target genes in the adult female mosquitoes. Insect Biochem. Mol. Biol. 40, 23–29.

Zhu, G.-H., Jiao, Y., Chereddy, S.C.R.R., Noh, M.Y., Palli, S.R., 2019. Knockout of juvenile hormone receptor, Methoprene-tolerant, induces black larval phenotype in the yellow fever mosquito, *Aedes aegypti*. Proc. Natl. Acad. Sci. U. S. A. 116, 21501–21507.

Zou, Z., Saha, T.T., Roy, S., Shin, S.W., Backman, T.W.H., Girke, T., White, K.P., Raikhel, A.S., 2013. Juvenile hormone and its receptor, methoprene-tolerant, control the dynamics of mosquito gene expression. Proc. Natl. Acad. Sci. U. S. A. 110, E2173–E2181.

CHAPTER THREE

The epitracheal endocrine system and associated signalling cascades in development, reproduction, and behaviour

Michael E. Adams[a,b]
[a]Department of Entomology, University of California, Riverside, CA, United States
[b]Department of Molecular, Cell, and Systems Biology, University of California, Riverside, CA, United States

Contents

Abstract

The epitracheal endocrine system is a network of glands whose functions during larval development are critical for initiation and scheduling of the ecdysis sequence at the end of each moult in insects. Ecdysis triggering hormones (ETHs) released from Inka cells recruit a cascade of downstream chemical signaling events that coordinate tracheal

Advances in Insect Physiology, Volume 60
ISSN 0065-2806
https://doi.org/10.1016/bs.aiip.2021.05.001

airfilling and sequential behaviours. Genes encoding ETHs and corresponding receptors are induced early in the molt by ecdysteroids, while Inka cell secretory competence depends upon late gene expression during steroid decline. During adulthood, ETH signaling is repurposed for reproductive functions, including promotion of vitellogenesis and ovulation in females and courtship memory in males. Some of these physiological actions depend upon ETH acting as a juvenile hormone promoting allatotropin. Emerging functions for ETH signaling include regulation of water balance, stress responses, and olfactory-mediated behaviours.

1. Introduction

The epitracheal endocrine system is a unique population of glands composed of both endocrine and exocrine cells situated on the respiratory (tracheal) system of insects. First described early in the 20th century as a collection of endocrine cells in the silkworm that fluctuate in size periodically (Ikeda, 1913), it was not until almost 80 years later that these secretory cells once again attracted attention through more detailed immunochemical and ultrastructural descriptions in the silkworm, gypsy moth, and waxmoth (Akai, 1992; Klein et al., 1999; Zitnan, 1989). Dusan Zitnan recognized the peptidergic nature of the largest cell in epitracheal glands (EG), which became known as the "Inka cell." Zitnan also noticed the apparent disappearance of Inka cells following ecdysis (Zitnan, 1989). He demonstrated a functional connection between these cells and ecdysis by injecting extracts of the EG into the hemocoel of pharate *Manduca* larvae, which led to pre-ecdysis and subsequent ecdysis behaviours within minutes. Further investigations of EGs in *Manduca* revealed Inka cells as the source of ecdysis triggering hormone (ETH), a 26-amino acid, C-terminally amidated peptide (Zitnan et al., 1996). Inka cells, which upon further scrutiny were shown to express multiple ETHs, since have been characterized widely in the Insecta (Roller et al., 2010; Zitnan and Adams, 2012; Zitnan et al., 2003).

The first descriptions of ETH signalling in the fruit fly *Drosophila* by Yoonseong Park ushered in a new phase of ecdysis analysis via genetic approaches (Park et al., 1999). Soon it became evident that secretion of ETHs from Inka cells is necessary and sufficient for successful completion of the moult by triggering ecdysis behaviours in *Drosophila* (Park et al., 2002). Release of ETH orchestrates a downstream signalling cascade in the central and peripheral nervous systems initiated by ETH acting as a "command chemical" (Kim et al., 2006a,b). However, more recent studies demonstrated interdependence of peptidergic signals involving eclosion hormone (EH), ETH, kinin, CCAP, and bursicon for orchestrating and scheduling ecdysis and related events, including tracheal airfilling, water balance, and wing expansion. This interdependence involves feedback and feedforward signalling loops involving multiple peptides necessary for neuronal responsiveness during the ecdysis signalling cascade (Krüger et al., 2015; Lahr et al., 2012; Mena et al., 2016). Furthermore, implementation of sophisticated genetic tools useful in the fruit fly *Drosophila* have revealed distinctive roles of ETH receptor splice variants ETHR-A and ETHR-B in scheduling of the ecdysis sequence (Diao et al., 2015a,b, 2017). Another noteworthy advance involves the discovery that EH is expressed more widely in the CNS than previously assumed, and like ETH is obligatory for ecdysis (Krüger et al., 2015; Mena et al., 2016; Scott et al., 2020).

One of the ongoing questions since identification of Inka cells in *Drosophila* has been the functional significance of Inka cell persistence into the adult stage (Park et al., 1999, 2002). Recent studies show that ETH signalling is re-purposed in adult *Drosophila* and other species for regulation of reproductive physiology and behaviour (Deshpande et al., 2018; Lee and Adams, 2021; Lee et al., 2017; Meiselman et al., 2017; Shen et al., 2021; Shi et al., 2019).

This review summarizes recent advances in our understanding of the functional significance of the epitracheal endocrine system and associated upstream and downstream signalling cascades in development and reproduction over the past 10 years and since the last comprehensive review on this subject (Zitnan and Adams, 2012). In the meantime, excellent reviews reporting on some of these advances in myriad and broader contexts have been published (Nässel and Zandawala, 2019, 2020; Taghert and Nitabach, 2012; White and Ewer, 2014).

2. Epitracheal glands, inka cells, and ETH signalling

In lepidopteran larvae, where they are best described, epitracheal glands are multicellular, consisting of the Inka cell, a smaller endocrine cell, an exocrine cell, and a duct cell, as described in a detailed ultrastructural study of the gypsy moth, *Lymantria dispar* and a subsequent study on *Manduca* (Klein et al., 1999; Zitnanová et al., 2001). Post-metamorphosis, Inka cells persist into the adult stage, but it appears that among the EG cells, they are the sole survivor.

EGs are prominent in freshly dissected pharate larvae and pupae by virtue of the Tyndall effect, in which densely packed secretory granules in the large

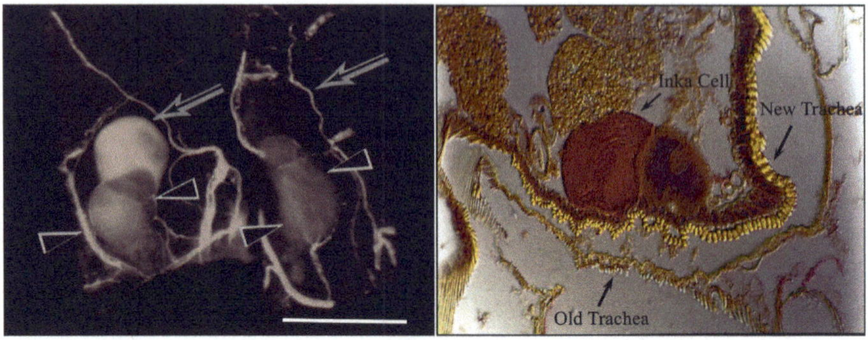

Fig. 1 (Left) Epitracheal gland of a pharate pupal *Manduca sexta* 3 h prior to ecdysis. The Inka cell (left arrow) exhibits the whitish, opaque appearance characteristic of the Tyndall effect. Following ecdysis, the Inka cell has released its secretory contents, thus losing its prominence and appears to disappear (right arrow). Two associated gland cells are visible on the basis of differential opacity (arrowheads). (Right) Inka cell visualized by immunohistochemical staining together with an associated endocrine cell following apolysis. Old tracheal lining and new trachea are indicated with arrows. *Left photo is from Zitnan, D., Kingan, T.G., Hermesman, J.L., Adams, M.E., 1996. Identification of ecdysis-triggering hormone from an epitracheal endocrine system. Science (New York, N.Y.) 271, 88–91. https://doi.org/10.1126/science.271.5245.88. Reprinted with permission from AAAS. Right photo by Dusan Zitnan.*

Inka cell produce pronounced light scattering, making them easily recognizable as bluish-white cells in live, unstained preparations (Fig. 1) (Zitnan et al., 1996). This property of the Inka cell along with immunohistochemical staining is responsible for its episodic prominence under the influence of high ecdysteroid levels during the moult and ecdysis-associated "disappearance" at moult culmination, when massive release and depletion of its secretory contents occurs (Zitnanová et al., 2001). Inka cells and ETH signalling molecules (ligands and receptors) are impressively cosmopolitan in the Insecta and also occur in crustaceans and arachnids (Roller et al., 2010; Zhu et al., 2019; Zitnan et al., 2003). A recent account showing occurrence of ETH in the Protura, a basal group within the Endognatha, indicates an early evolutionary origin (Christie and Chi, 2015).

In the fruit fly *Drosophila*, Inka cells can be visualized via expression of green fluorescent protein (GFPs) reporters in transgenic animals (Fig. 2), but next to nothing is known about associated epitracheal gland cells (Park et al., 2002). Loss of fluorescence reporter from Inka cells of transgenic flies has provided a means for accurate timing of ETH release at onset of the ecdysis sequence (Kim et al., 2006b). Other methods for measuring ETH release include loss of immunohistochemical staining and enzyme

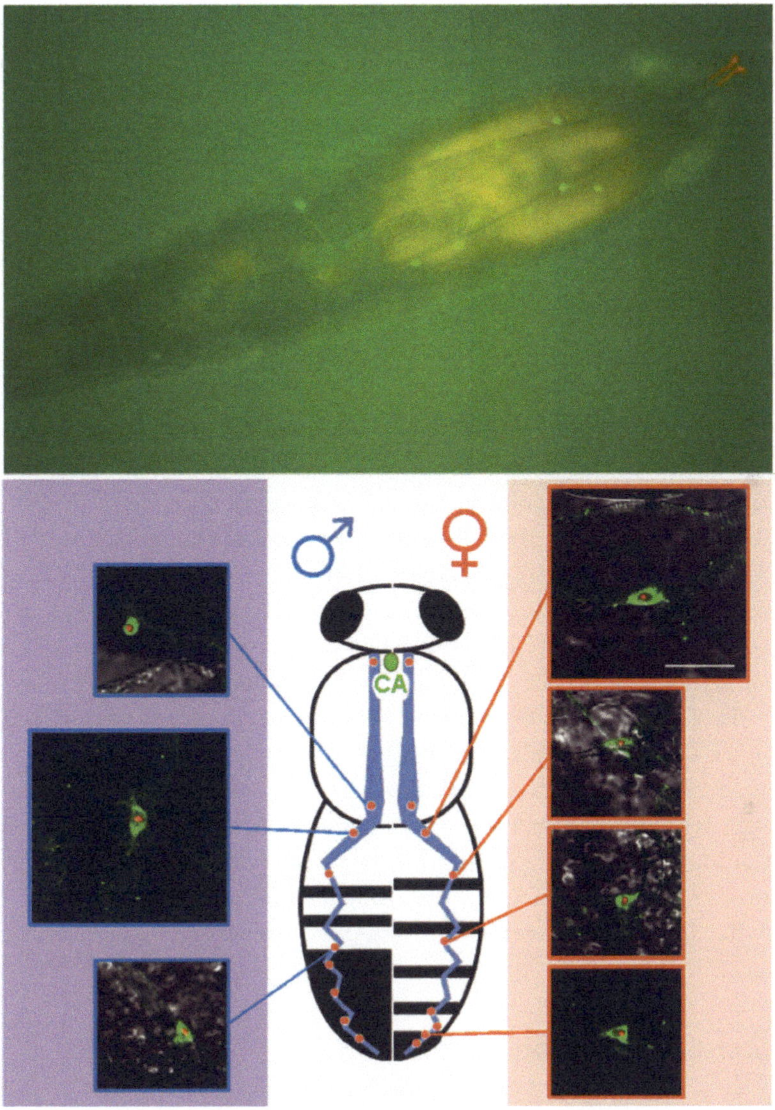

Fig. 2 Inka cells of *Drosophila*. Upper: Inka cells in a transgenic 3rd instar larva visualized via GFP reporter. Lower: Inka cells in adult Drosophila double-labelled with red-stinger localized in the nucleus and immunohistochemical staining (green) of cytoplasm. Upper photo by Do-Hyoung Kim. *Lower picture taken from Meiselman, M., Lee, S.-S., Tran, R.-T., Dai, H., Ding, Y., Rivera-Perez, C., Wijesekera, T.P., Dauwalder, B., Noriega, F.G., Adams, M.E., 2017. Endocrine network essential for reproductive success in Drosophila melanogaster. Proc. Natl. Acad. Sci. U. S. A. 114, E3849–E3858. https://doi.org/10.1073/pnas.1620760114.*

immunoassay (Clark et al., 2004; Ewer et al., 1997; Zitnan et al., 1999). Use of enzyme immunoassays enabled quantification of circulating Inka cell peptides during ecdysis behaviours in pharate *Manduca* larvae. Peak concentrations of these peptides and some of their precursor proteins reach 20–40 nanomolar during the performance of the ecdysis sequence, placing them squarely in the mid- to upper range of their concentration-response curves for receptor activation (Kim et al., 2006a; Zitnan et al., 1999).

2.1 Inka cell peptides

In the lepidopterous insects *Bombyx* and *Manduca*, three peptides have been characterized: PETH, ETH, and ETH-AP, while in *Drosophila* two peptides called ETH1 and ETH2 have been characterized. ETH-AP is an "associated peptide" for which no discernable function has been assigned. While some division of labour has been observed for these peptides, overall there is considerable redundancy in their functions. For example, injection of PETH into *Manduca* larvae elicits preecdysis I behaviour only, while ETH elicits all three behavioural phases: preecdysis I, preecdysis II, and ecdysis. In contrast, either PETH or ETH is capable of recruiting all three phases of the behavioural sequence in the silkworm, *Bombyx* (Zitnan et al., 1999, 2002). In *Drosophila*, ETH1 elicits all phases of preecdysis and ecdysis, whereas ETH2 fails to elicit preecdysis, but triggers ecdysis (Park et al., 2002). Interestingly, either ETH1 or ETH2 is sufficient to elicit tracheal airfilling, which precedes pre-ecdysis behaviour. While endocrine signalling leading to tracheal airfilling has been investigated in *Drosophila* (see Section 3.5 for further details), the story in this regard for other insects is somewhat limited and well summarized in a recent review (Förster and Woods, 2012).

2.2 Steroid regulation of inka cells

Inka cells are multinucleate and increase in volume in response to the surge of ecdysone that initiates cessation of feeding and initiation of the moult. This volume increase is associated with robust transcription of the ETH precursor gene, whose upstream promoter contains ecdysone response elements (EcREs) in both moths and flies (Park et al., 1999; Zitnan et al., 1999). Ecdysone receptor isoforms EcR-B1 and EcR-B2 are expressed in the Inka cell (Gauthier et al., 2012; Zitnan et al., 1999). Gauthier et al. demonstrated that B2 expression is under specific regulation of the coactivator protein Cryptocephal (CRC), a *Drosophila* homologue of the mammalian activating transcription factor 4 (ATF4), a basic leucine-zipper protein known to be involved in stress responses. Interestingly, in addition to its well-known morphogenetic actions during development, ecdysone appears to play diverse roles in stress responses in a manner similar to glucocorticoids in vertebrates (Gruntenko et al., 2010; Ishimoto et al., 2009; Meiselman et al., 2018).

Similar to studies in *Manduca*, ETH expression in *Drosophila* Inka cells is promoted by ecdysone. On the other hand, CRC mutants exhibit reduced ETH expression. Using the ETH-GeneSwitch-Gal4 driver to conditionally express dominant negative EcR mutant isoforms specifically in the Inka cell, it was demonstrated that resulting ecdysis mutant phenotypes (double mouth hooks, lethal ecdysis deficiency) were partially rescued by EcR-B1 but fully rescued by EcR-B2 expression (Gauthier et al., 2012). This study confirms the necessary role of ecdysone and multiple EcR isoforms in regulation of ETH expression in the Inka cell. It also provides evidence that different EcR isoforms together with diverse coactivators may regulate cellular processes associated with ETH transcription, accumulation, and secretory competence independently. As such, the Inka cell provides a unique model for further investigation of how ecdysone regulates diverse cellular and stage-specific processes.

While high levels of ecdysone promote expression of the ETH gene, they inhibit its release by delaying acquisition of Inka cell secretory competence, with decisive consequences for precisely scheduled behavioural events at moult termination (Kingan and Adams, 2000; Zitnan et al., 1999). In particular, delaying ETH release until late stages of the moult is protective. This was demonstrated by injection of ETH at various times during later stages of the moult when steroid levels are still relatively high. Such ETH injection leads to premature ecdysis behaviour, but unsuccessful shedding of the cuticle, since digestion and recycling of old cuticle constituents is incomplete (Zitnanová et al., 2001). Activation of ETH receptors in CNS target neurons causes them to be internalized and, if steroid levels remain somewhat elevated but insufficient to induce receptor re-expression and insertion into the cell membrane, subsequent natural release of ETH does not trigger the behaviours necessary for cuticle shedding. Thus, inability to release ETH until the animal is absolutely ready for cuticle shedding can be viewed as a "fail-safe" protective measure preventing premature ecdysis and ensuring survival to the next stage of development.

Given the importance assigned to precise timing of Inka cell secretory competence, what cellular events are involved in acquisition of secretory competence? First of all, steroid levels must decline to quite low levels (Kingan and Adams, 2000). In *Manduca* pharate pupae, this occurs 8–10 h prior to ecdysis, when steroid levels fall below ~0.1 μg/mL. In *Drosophila* larvae, expression of the "competence factor" βFTZ-F1 occurs late in the moult, coincident with declining steroid levels (Yamada et al., 2000). Yamada et al. showed that βFTZ-F1 mutants, while progressing through formation of new cuticle, spiracles, mouthparts and trachea, fail to exhibit ecdysis phenotypes, including air bubble translocation in the pharate pupal stage, head eversion, and elongation of legs and wings.

A recent study demonstrated that RNAi knockdown of *βftz-f1* transcripts selectively in Inka cells causes lethal ecdysis deficiencies (Cho et al., 2013). In moulting larvae, these were accompanied by failure of Inka cells to release ETH at the appropriate time and the "buttoned-up" phenotype described years earlier following deletion of the *eth* gene (Park et al., 2002). Significantly, most all of the *βftz-f1* mutant phenotypes described earlier by Yamada et al. were duplicated by Inka cell-specific knockdown of this gene. Furthermore, Inka cell *βftz-f1* knockdown-related ecdysis deficiency phenotypes were observed at all stages of development: larval, pupal, and adult ecdyses. Finally, overexpression of *βftz-f1* specifically in Inka cells rescued most of the phenotypes of organism-wide $\beta ftz\text{-}f1^{ex7}$ deletion mutant flies. These results confirmed that the major defects arising from $\beta ftz\text{-}f1^{ex7}$ flies described earlier by Yamada et al. can be assigned to ETH deficiency and confirm the primary role this gene in conferring secretory competence on Inka cells during the late stages of the moult. Inka cells therefore can be considered as obligatory "gateway" cells through which development proceeds in late stages of each moult.

2.3 ETH receptors

ETH signalling operates through two subtypes of G protein-coupled receptors: ETHR-A and ETHR-B (Iversen et al., 2002; Park et al., 2003). These receptors arising from the same gene are alternative splicing isoforms involving 3-prime 4a and 4b exons (Iversen et al., 2002; Kim et al., 2006a; Park et al., 2003). Neurons expressing these receptors mobilize calcium from intracellular stores, leading to the supposition that they are G_q-coupled. However, a recent study, the first of its kind involving the hemimetabolous locust *Schistocerca gregaria*, showed that receptor activation can result in dual responses involving calcium mobilization and cyclic AMP elevation (Lenaerts et al., 2017).

Early accounts of CNS expression indicate that ETHR-A and ETHR-B expression occurs in largely mutually exclusive neuronal populations with a few exceptions, including EH-expressing Vm neurons and posterior-dorsolateral NPF-expressing neurons in *Manduca* (Diao et al., 2015a; Kim et al., 2006a,b). By far, the predominantly peptidergic ETHR-A neurons are the best characterized and are involved in all phases of the ecdysis sequence. While a subset of ETHR-B neurons also is peptidergic, many appear to be those primarily signalling via classical transmitters, including acetylcholine and glutamate (Diao et al., 2015a).

Mirroring the temporal pattern of ETH gene expression, ETHR expression is episodic and dependent upon rising ecdysteroid levels. Zitnan showed early on that behavioural sensitivity to ETH along with its ability to elicit fictive behaviour occurs around the time of apolysis when ecdysteroid levels peak (Zitnan et al., 1999). Loss of ETH sensitivity post-ecdysis is correlated with disappearance of receptor transcripts in target neurons and likely internalization of liganded receptor proteins. ETH sensitivity is recovered at the next moult or can be induced by artificial steroid exposure (Zitnanová et al., 2001). Periodic, steroid-dependent ecdysis behaviours therefore can be viewed as excellent paradigms for detailed investigations of the relationship between gene expression and encoding of innate behaviours.

Recent innovations of genetic approaches to ecdysis analysis have greatly facilitated assignment of functional roles for ETHR-A and ETHR-B receptor splice isoforms (Diao et al., 2015a,b). One of these is the "plug-and-play" strategy utilizing "Trojan exon" cassette insertion into the ETHR via splice acceptor sequences followed by a viral T2A ribosome skipping element for Gal4 incorporation selectively into ETHR-A or ETHR-B neurons. This strategy allows for high fidelity labelling of the two receptor isoforms and manipulation of their expression under control of the endogenous ETHR promoter.

Results of these studies show that, as demonstrated earlier using traditional *in situ* hybridization approaches (Kim et al., 2006a,b), ETHR-A and ETHR-B isoforms are expressed largely in separate populations of neurons. Diao et al. showed that ETHR-A neurons are obligatory for all ecdyses—larval, pupal and adult. Suppression of ETHR-B expression by electrical silencing (Kir expression) or interference with splice insertion was found to be inconsequential for larval ecdysis, but expression of this isoform is obligatory for pupal and adult ecdyses (Diao et al., 2015a). Furthermore, ETHR-B silencing inhibits pupal preecdysis (Phase I) behaviour, but not ecdysis (Phase II) or postecdysis (Phase III) behaviours (Diao et al., 2017).

The obligatory nature of ETH signalling in triggering ecdysis was first demonstrated by Park, who deleted the gene encoding the peptide precursor through excisional mutagenesis (Park et al., 2002). Since then, the essential role of ETH signalling in *Drosophila* ecdysis has been validated by others and demonstrated to operate in similar fashion in the silkworm *Bombyx* (Dai et al., 2008), oriental fruit fly *Bactrocera* (Shi et al., 2017), and Colorado potato beetle *Leptinotarsa* (Shen et al., 2020).

3. Initiation and scheduling of the ecdysis sequence

3.1 Two modes of ETH release from inka cells

The endocrine basis for insect ecdysis/eclosion was demonstrated by James Truman in a series of groundbreaking publications during the latter part of the 20th century describing eclosion hormone. This peptide hormone was eventually characterized as a 62-amino acid, internally disulphide-bonded peptide produced by two pairs of brain neurons in lepidopterous insects (Kataoka et al., 1987; Marti et al., 1987; Truman, 1992a). Eclosion hormone induces ecdysis behaviour in whole animals upon injection and fictive ecdysis behaviour upon exposure of the central nervous system. An important caveat to the latter action of eclosion hormone in these experiments was the requirement for preparation of the isolated nervous system together with an intact tracheal system. The assumption was that the highly aerobic CNS could not produce fictive eclosion behaviour without a sufficient air supply. In fact, the requirement of an intact tracheal system was later determined instead to be related to the presence of epitracheal Inka cells.

The next stage of the ecdysis/eclosion story arose through studies by Dusan Zitnan, who became interested in a set of paired, segmentally repeated endocrine cells observed in waxmoth larvae (Zitnan, 1989). He found that an antiserum raised against a molluscan peptide called SCP_B stains these cells immunohistochemically. In the mid-1990s, he demonstrated that extracts of just a few of these cells triggers ecdysis just as eclosion hormone does. One of the active ingredients of Inka cells turned out to be a 26-amino acid peptide with a C-terminal amino acid sequence motif similar to that of SCP_B and other "pyrokinin" peptides (Zitnan et al., 1996). Upon injection of this peptide, named *Manduca sexta* ecdysis triggering hormone (MasETH), larvae respond with preecdysis following by ecdysis behaviour. Soon thereafter, *Bombyx* and *Drosophila* ETHs were identified, which trigger ecdysis behaviours indistinguishable from naturally occurring events (Adams and Zitnan, 1997; Park et al., 1999).

This was a surprise, since the hormonal basis for ecdysis had been shown already years earlier to involve eclosion hormone at all larval, pupal, and adult transitions (EH) (Truman, 1992b). This apparent conundrum was explained by demonstration that EH serves as an ETH releasing factor, acting on a receptor guanylyl cyclase expressed in Inka cells to cause ETH release (Chang et al., 2009; Kingan et al., 1997; Zitnan et al., 1996). ETH in turn acts directly on the central nervous system to trigger the ecdysis behavioural sequence. In contrast, exposure of the isolated, trachea-less CNS to EH fails to trigger fictive ecdysis unless Inka cells are included in the bathing medium (Zitnan et al., 1996).

The role of EH as a releasing factor was soon revealed to be just part of the story, however, when Ewer and Truman showed that ETH also functions an EH releasing factor. Indeed, the two peptides engage in a positive feedback signalling loop, whereby ETH release from Inka cells reinforces EH release (Ewer et al., 1997). Subsequent studies described below have shown that both ETH and EH along with a plethora of downstream signalling molecules act on the CNS to orchestrate the behavioural and physiological events leading to ecdysis.

These observations led to the obvious question regarding upstream signals for ETH release. Zitnan took this one step further by injecting the neuropeptide corazonin into *Manduca* larvae, anticipating its well-known action promoting cuticle melanization (Tanaka, 2006). Unexpectedly, the result was initiation of pre-ecdysis behaviour. This led to further scrutiny of the early endocrine steps in initiation of the ecdysis sequence and demonstration that, at least in *Manduca*, corazonin appears in the hemolymph prior to pre-ecdysis and functions as an initiator of early steps in the ecdysis sequence (Kim et al., 2004).

Corazonin is expressed in medial neurosecretory cells $1a_1$ known to be influenced by the circadian clock, thus providing insight into the timing of ecdysis and eclosion (Truman, 1992a). Furthermore, the realization that both corazonin and eclosion hormone are ETH releasing factors provided incentive for a more detailed study of their relative roles in release. It became clear that two modes of ETH release are controlled by these neuropeptides: an early low-level release caused by corazonin and subsequent massive release of ETH caused by EH-induced elevation of cGMP. The early low level release of ETH involves elevation of intracellular calcium via Gq-coupled corazonin receptor activation. Interestingly, corazonin-induced ETH release appears to be limited by rapid desensitization of the corazonin receptor. This low-level ETH release leads to activation of EH neurons.

EH release in turn activates EH receptors in Inka cells, which are receptor guanylyl cyclases in the oriental fruit fly, *Bactrocera* (Chang et al., 2009). Elevation of cGMP in Inka cells causes massive ETH release and transition from pre-ecdysis to ecdysis. EH receptor homologues have yet to be characterized in other insects.

The role of corazonin in early release of ETH from Inka cells has been documented only in lepidopterous insects; it does not appear not to function as such in *Drosophila*. And while eclosion hormone is sufficient in causing ETH release from Inka cells in *Drosophila*, ablation of Vm neurons does not preclude ETH release or change the timing of preecdysis and ecdysis scheduling (Clark et al., 2004). It is likely that redundant mechanisms contribute to ETH release, even in the absence of corazonin and Vm neurons. For example, decline of ecdysone to low levels late in the moult may be sufficient to cause spontaneous ETH release. It also has been reported recently that EH is expressed in an expanded ensemble of neurons that may compensate for loss of brain Vm neurons (Scott et al., 2020).

3.2 Activation of CNS ETHR peptidergic ensembles and ecdysis scheduling

Shortly after discovery of ETH and the epitracheal endocrine system (Zitnan et al., 1996), Gammie and Truman reported studies with *Manduca* indicating that ETH elicits preecdysis "directly," but causes ecdysis indirectly via a signalling cascade involving ETH→EH→CCAP (Gammie and Truman, 1997, 1999). Their studies also indicated that CCAP both activates ecdysis and inhibits preecdysis. These insights were prescient in the context of more recent studies using genetic tools in *Drosophila*, even as the complexity of the ecdysis signalling cascade has grown considerably.

Several technical advances allowed for identification and monitoring of ETH target neurons in the CNS and their association with events in the ecdysis sequence, including tracheal air filling, preecdysis, ecdysis, and postecdysis. First, studies in *Drosophila* led to identification of ETH receptors, facilitating mapping of target neurons, many of which are peptidergic (Iversen et al., 2002; Kim et al., 2006a,b; Park et al., 2003). These central peptidergic ensembles express kinin, FMRFa, CCAP, MIPs, ASTCC, CRF-like diuretic hormones (DH), bursicon, and likely additional peptides. Soon thereafter, use of their peptide gene promoters for creation of Gal4 drivers facilitated cell-specific calcium imaging of ETHR neuron activity in *Drosophila*. Studies by Young-Joon Kim and Do-Hyoung Kim (Kim et al., 2006b, 2015) demonstrated sequential calcium mobilization first in FMRFa and kinin neurons and later in ensembles of neurons expressing CCAP.

Identification of ETHR neurons in *Drosophila* facilitated further studies of homologous peptidergic ensembles in *Manduca*, contributing to a "cross-fertilization" of knowledge using these two organisms. *Manduca* lent itself to electrophysiological experiments (Kim et al., 2006a), whereas *Drosophila* offered the power of genetic tools, including RNA interference, receptor overexpression, and calcium imaging as a measure of neuronal activity (Diao et al., 2017; Kim et al., 2006b, 2015; Mena et al., 2016).

In *Manduca*, L3,4 neurons expressing kinins and CRF-like diuretic hormones (DH) were associated with initiation of preecdysis behaviour, while IN704 neurons expressing CCAP/ASTCC/bursicon/MIP were associated with ecdysis behaviour. These hypotheses were supported by experiments in which application of these "downstream" peptide cocktails to isolated segmental nerve cords of *Manduca* larvae elicited fictive behaviours identical to those produced by exposure to PETH and ETH (Kim et al., 2006a). Interestingly, while actions of PETH and ETH lead to expression of the entire ecdysis sequence upon initiation, as is characteristic of a classical "fixed action pattern," application of downstream signalling molecules have reversible effects on behaviour. Furthermore, Zitnan was able to show that bath application of the peptide cocktails expressed in L3,4 and IN704 neurons produced higher fidelity, more robust behaviours then any of the individual peptides. These physiological experiments conducted in *Manduca* and *Bombyx* constitute a "similar action" argument implicating these downstream ETHR neuron peptides in orchestration of hormonally-induced sequential behaviours (Kim et al., 2006a; Zitnan et al., 1999, 2002). A recently published account by Daubnerova et al. provides an expanded and deeper understanding of ETH receptor networks operating in *Bombyx*, both in the CNS and periphery, including frontal ganglion, H-organ, gut, epidermis, and Malpighian tubules, implicating ETH signaling in a wide variety of physiological functions (Daubnerova et al., 2021).

The "similar action" argument developed in *Manduca* implicating activity in kinin/DH neurons in preecdysis initiation and CCAP/ASTCC/MIP/bursicon (CAMB) neuron activity with ecdysis was tested further in *Drosophila* using calcium imaging of neuronal activity on the isolated CNS to correlate timing of neuronal activity with behaviour (Kim et al., 2006b, 2015). An advantage of this paradigm is that ecdysis behaviours are centrally patterned according to a schedule that is preserved with high fidelity whether measured as fictive activity in the isolated CNS or as behaviour in the intact animal. On this basis, the ecdysis sequence can be fairly viewed as a classical fixed action pattern, which runs to completion once initiated without sensory input to maintain the basic temporal and physical attributes of the sequence (Schleidt, 1974). An exception to this is that ecdysis termination is dependent upon loss of sensory input once the cuticle has been physically shed; in its absence, fictive behaviour is prolonged for a considerable time (Kim et al., 2015; Zitnan and Adams, 2000).

Via calcium imaging measurements, Young-Joon Kim and Do-Hyoung Kim demonstrated that FMRFa and kinin neurons become active within minutes of ETH exposure, coincident with early events including tracheal airfilling and preecdysis behaviour (Fig. 3). The delayed onset of CAMB neuron activity is associated with onset of ecdysis and postecdysis behaviours

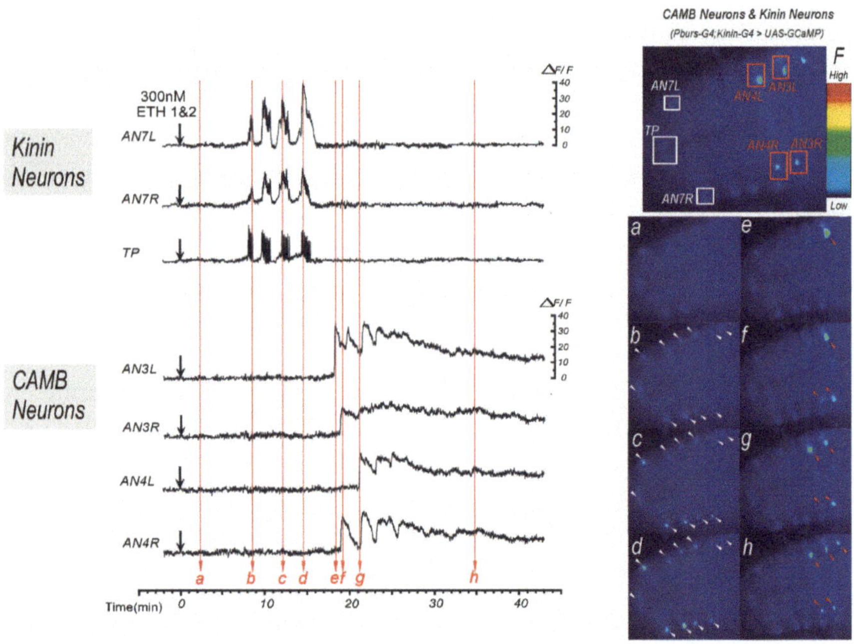

Fig. 3 ETH evokes sequential activation of kinin and CAMB neurons during the pupal ecdysis sequence as visualized by calcium imaging. Application of ETH1 and ETH2, each at 300nM concentration (downward arrows) leads to onset of calcium mobilization in Kinin neurons beginning at 8.5min. Downward red arrows labelled with lower case letters (a–h) refer to images at the lower right depicting activation of denoted cells at various times after ETH application. Following cessation of kinin neuron activity, CAMB (CCAP/ASTCC/MIP/Bursicon) neurons become active after a short delay. *Taken from Kim, D.-H., Han, M.-R., Lee, G., Lee, S.-S., Kim, Y.-J., Adams, M.E., 2015. Rescheduling behavioral subunits of a fixed action pattern by genetic manipulation of Peptidergic signaling. PLoS Genet. 11, e1005513. https://doi.org/10.1371/journal.pgen.1005513.*

(Kim et al., 2006b, 2015). Indeed, activation of CAMB neurons using temperature-sensitive Trp-M8 expression was shown to initiate the ecdysis motor pattern.

Similar results associating activity of CCAP-ETHR-A neurons with ecdysis behaviour were obtained by John Ewer, Benjamin White, and colleagues (Diao et al., 2017; Mena et al., 2016). Further testing to establish causality was accomplished through expression of temperature-sensitive Trp-A1 ion channels in ETHR-A neurons, allowing for selective activation through temperature manipulation. Use of these methods demonstrated that activation of CAMB neurons causes initiation of Phase II/ecdysis behaviour and Phase III/postecdysis behaviours (Diao et al., 2017). Indeed, activation of CCAP-ETHR-A neurons was sufficient to inhibit preecdysis behaviour, reminiscent of the results reported by Gammie and Truman in *Manduca* (Gammie and Truman, 1997). The role of CAMB neurons, which co-express bursicon, is in agreement with an earlier report demonstrating that bursicon is a dominant proximal trigger for ecdysis onset, with CCAP contributing to robustness of behavioural expression (Kim et al., 2015; Lahr et al., 2012; Mena et al., 2016).

Another approach useful in connecting peptidergic neurons with sequential ecdysis behaviours was taken by Kim et al. (2015). This study showed that titrating receptor expression in ETHR peptidergic ensembles could provide clues to their involvement in each of the sequential ecdysis behaviours. For example, overexpression of ETHRs in kinin neurons prolongs preecdysis duration, whereas RNAi knockdown of ETHRs reduces its duration. On the other hand, overexpression of ETHR in CAMB neurons accelerates the switch from preecdysis to ecdysis, presumably by increasing sensitivity of these neurons to ETH. ETHR knockdown in these neurons has the opposite effect. These findings provide further support for involvement of kinin neurons with preecdysis behaviour and CAMB neurons with ecdysis behaviour.

Nevertheless, activation of *Drosophila* kinin neurons using either TrpM8 or TrpA1 failed to initiate pre-ecdysis, even though disruption of kinin signalling led to increased variability in onset and duration of preecdysis (Kim et al., 2015). Therefore, while kinin neurons may be involved in scheduling preecdysis, their activation alone is insufficient to trigger the behaviour. Further insight into the issue of preecdysis regulation is provided by observations that electrical silencing of ETHR-B neurons blocks Phase I/preecdysis behaviour (Diao et al., 2017). Specific identities of these ETHR-B neurons remain undefined.

3.3 Nature of the delay of ecdysis onset

One of the ongoing questions raised by these experiments involves the cellular basis for the delay in onset of ecdysis behaviour once the CNS has been exposed to ETH. This delay was assigned to inhibition provided by neurons descending from the subesophageal (SG) and thoracic ganglia (TG) (Fuse and Truman, 2002; Zitnan and Adams, 2000). Ligatures applied at various intervals after exposure of *Manduca* larvae to ETH showed that, while ecdysis circuitry in abdominal ganglia were "activated" within 5–10 min of ETH exposure ("activation" defined as elevation of cGMP levels in ETHR neurons), onset of ecdysis behaviour was delayed 40–60 min. This delay could be reduced or eliminated by separation of thorax from abdomen by ligation or by severing the connection between subesophageal (SG), thoracic (TG) and abdominal ganglia (AG). The precise source of this inhibition has yet to be identified, but Zitnan observed cGMP elevation in SG and TG neurons even hours before ETH release and also noted that immunohistochemical staining intensity increased further following ETH release. These neuronal elements are therefore implicated in the descending inhibition responsible for ecdysis delay. Recent studies in *Drosophila* provide additional insight relevant to this issue.

Further to the mechanism(s) of descending inhibition underlying the delay in ecdysis onset (Baker et al., 1999; Zitnan and Adams, 2000), Kim et al. recently demonstrated that expression of the $G_{i/o}$ inhibitor pertussis toxin in CAMB neurons accelerates the switch from preecdysis to ecdysis (Kim et al., 2015). Furthermore, silencing non-CCAP MIP neurons accelerates the switch from preecdysis to ecdysis. This implicates descending inputs to CAMB neurons such as peptidergic MIP neurons that engage in inhibitory Gi/o signalling. Given that ETHR-B neurons in *Drosophila* are ~450 × more sensitive to ETH than are ETHR-A neurons (Park et al., 2003), one hypothesis that follows from the Diao et al. result showing that silencing ETHR-B blocks preecdysis is that descending inhibitory ETHR-B neurons (MIP neurons?) activated early in the ecdysis sequence provide the descending inhibition that delays ecdysis onset. As ETH levels rise, excitatory input to CAMB neurons overcomes this inhibition to initiate ecdysis. Further studies on this subject are needed clarify this mechanism.

The delay in switching from preecdysis to ecdysis appears also to involve inhibitory influences of the neurotransmitter GABA (Mena et al., 2016). In this study, application of the GABA-A receptor blocker picrotoxin shortened the latency to ecdysis as did RNAi knockdown of GABA-A receptors in CCAP-ETHR-A neurons. These findings provide additional complexity to putative circuit components that programme the fixed intervals inherent in the ecdysis sequence.

3.4 Necessity and sufficiency of modulatory signals for the ecdysis sequence

Since the early days of studies on the relationship between EH and ETH, our understanding of their necessary and sufficient roles in orchestration of the ecdysis sequence has gone through successive phases of evolution. Clearly, the elegant use of genetic manipulations by John Ewer and colleagues working with *Drosophila* has been instrumental in advancing our understanding. Early studies on the role of EH that made use of cell death genes to kill the EH-expressing Vm neurons led to the conclusion that this hormone plays a "significant but nonessential role in ecdysis" (McNabb et al., 1997). This has turned out to be quite an understatement. The implication was that ETH alone suffices as an ecdysis trigger. One of the first clues that EH plays a more critical role was a report by the Ewer lab showing that injection of ETH into flies lacking EH Vm neurons fails to initiate premature ecdysis, as it does in control flies (Clark et al., 2004). While this result hinted at the fact that at least some cellular responses to ETH are dependent on presence of EH, it was difficult to reconcile with earlier observations showing that ETH knockouts are lethal (Park et al., 2002), whereas EH "knockouts" through ablation of Vm neurons are inconsequential.

Next came a study from the same lab showing that EH null flies, in which the gene encoding EH is deleted, suffer lethal ecdysis deficiency (Krüger et al., 2015). EH-null mutants apparently fail to release ETH, lack preecdysis behaviour, and somehow fail to perform behaviours with sufficient efficiency to permit transition to the next stage via successful ecdysis. Injection of ETH does not rescue these animals, but ectopic expression of EH (in Inka cells!) of these mutant flies is 100% effective in rescuing ecdysis deficiency. This study confirms that, in *Drosophila*, influences of EH on the CNS and likely the periphery, are necessary for ETH release and also for competence of ETH target cells to respond (Krüger et al., 2015). These findings highlight another salient feature of ecdysis signalling: feed-forward conditioning.

Further work on the obligatory roles of EH was reported in a subsequent study from the Ewer lab, which probed more deeply deficiencies in EH-null mutants making extensive use of calcium imaging (GCaMP) and optical voltage recordings (ArcLightα). They found further evidence for feed-forward influences of EH on ETHR neurons, autocrine influences of bursicon released from CCAP/CAMB neurons, and influences of GABA on the timing of ecdysis onset. For example, EH deficiency causes temporal changes in calcium dynamics as well as a diminution in amplitudes of CCAP-ETHR-A responses to ETH. Similarly, flies hemizygous for the pburs gene essential for assembly of the heterodimer bursicon also showed reduced amplitudes of CCAP neuron calcium responses. It seems that ETH-induced release of bursicon from CAMB neurons leads to bursicon-induced "autocrine" conditioning of these same neurons to upstream modulators. Bursicon deficiencies translate into weaker and more disorganized ecdysis motor patterns. This study confirms essential roles for EH and bursicon in conditioning downstream targets of ETH in the CNS (Mena et al., 2016).

In the most comprehensive study of the ecdysis behavioural sequence to date, Diao et al. describe how "neuromodulatory connectivity" configures the pupal ecdysis sequence triggered by the "hormonal initiator" ETH via three neuronal "layers": (1) An input layer consisting of ETHRA-CCAP neurons, non-CCAP-ETHRA neurons, and ETHRB neurons, (2) a central pattern generator layer containing bursicon receptor (Rk) neurons, and (3) a motoneuron output layer consisting of CCAP-R (receptor) neurons consisting largely of motoneurons (Fig. 4) (Diao et al., 2017). The strength of this study lies in the use of powerful genetic tools involving intersectional Gal4 drivers for precise control of subpopulations of neuronal elements, remote temperature-mediated control of neuronal activity and incorporation of output muscle activity in the overall cellular activities under observation. These authors provide evidence that ETHR-B neurons are required for Phase I/preecdysis behaviour, while bursicon configures CPG activity for Phase II/ecdysis and Phase III/postecdysis behaviours. CCAP acts primarily on motoneurons for output signalling to the musculature.

These authors used remote neuronal activation tools including the ATP/P2X2 physiogenetic strategy (Yao et al., 2012) and Trp-A1 to drive bursicon receptor neurons (Rk) neurons to elicit phases II and III. However, driving of ETHR-B neurons implicated in Phase I failed to elicit the fictive behaviour. It seems possible that a combination of ETHR-A (e.g. possibly including kinin neurons) and as yet unidentified ETHR-B neurons are needed for activation of preecdysis/Phase I behaviour.

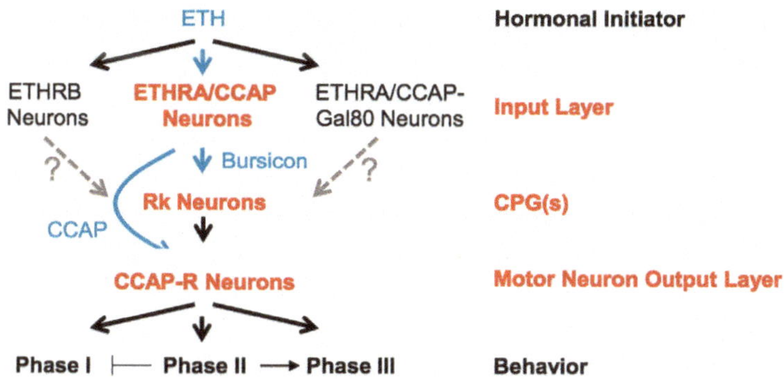

Fig. 4 Hierarchical organization of the pupal ecdysis circuit. Each layer in the network hierarchy (red) is the target of one of the ecdysis hormones (blue), as defined by expression of its receptor. The most important component of the input layer, ETHRA/CCAP, participates in regulating all three behavioural phases and is the source of CCAP and Bursicon. The Bursicon receptor Rk defines a central pattern generating layer, which sends output to a population of motor neurons that express the receptor for CCAP and are required for the ecdysis sequence. Solid arrows indicate demonstrated functional connections, while grey, dashed arrows indicate hypothesized connections. Although the detailed mechanisms governing motor programme progression remain to be determined, those that promote Phase II negatively regulate Phase I and positively regulate Phase III. *From Diao, Feici, Elliott, A.D., Diao, Fengqiu, Shah, S., White, B.H., 2017. Neuromodulatory connectivity defines the structure of a behavioral neural network. Elife 6, e29797. https://doi.org/10.7554/eLife.29797.020.*

3.5 Tracheal airfilling

Late in the moult, the tracheal system is filled with moulting fluid. Early events occurring in the ecdysis/eclosion sequence following ETH release include collapse of old tracheal tubes around which a new epidermis and cuticular layer have been constructed during the moult (Park et al., 2002). Subsequently, new tracheae are filled with air and old collapsed trachea are pulled out during ecdysis. Molecular events involved in tracheal airfilling remain obscure, but considerable evidence supports the notion that it is under endocrine control.

At adult eclosion, cGMP elevation is observed in trachea epithelial cells (Baker et al., 1999). This process is compromised permanently in EH-deficient animals, with concomitant failure of tracheal airfilling. This failure can be rescued by injection of the membrane-permeant 8-bromo-cGMP. These findings support a primary role of EH in tracheal airfilling that occurs during adult eclosion. ETH injection also triggers airfilling, likely via downstream EH release.

During larval ecdysis, collapse of old trachea and airfilling of new trachea occur within minutes of ETH release (Park et al., 2002). Deletion of the ETH gene delays tracheal collapse and airfilling, but does not eliminate it permanently. Occurrence of these events some hours later may be due to delayed EH release, but other modulators also may be involved. One candidate may be FMRFa, which is released early in the pupal ecdysis sequence (Kim et al., 2006b). FMRFa is a known ligand for DEG/ENaC ion channels expressed in tracheal epithelial cells (Assmann et al., 2014). Another neuropeptide implicated in tracheal airfilling is kinin, also referred to in many accounts as leucokinin or drosokinin (Kim et al., 2018).

In *Drosophila*, activity of kinin neurons occurs within minutes of ETH exposure, coincident with tracheal collapse and airfilling (Kim et al., 2015). Disruption of kinin signalling leads to defective airfilling during all larval stages (Kim et al., 2018). Such defects are observed upon ablation or electrical silencing of kinin neurons, as well as RNA silencing of the kinin gene or the ETH receptor in kinin neurons, indicating that ETH targets kinin neurons to promote tracheal airfilling. A hypomorphic kinin receptor mutant fly line (Lkr ^{f02594}), where kinin receptor expression is reduced ~25%, exhibits partial, but significant tracheal air-filling defects in all larval stages. Targeted Kinin receptor silencing in tracheal epithelial cells using breathless or pickpocket (ppk) drivers compromises tracheal air-filling. On the other hand, promotion of kinin signalling *in vivo* through peptide injection or kinin neuron activation via TrpA1 (dTrpA1) expression induces premature tracheal collapse and airfilling (Movie 1 in the online version at https://doi.org/10.1016/bs.aiip.2021.05.001). Interestingly, premature tracheal collapse and airfilling induced by Trp-A1 is reversible. Moreover, direct exposure of tracheal epithelial cells *in vitro* to kinin leads to calcium mobilization in tracheal epithelial cells. These findings firmly implicate kinin signalling in tracheal airfilling and support mechanisms involving ion and fluid transport across trachea epithelial cells, given the extensive literature on the diuretic actions of this peptide. An extensive and holistic review on tracheal airfilling can be found in a recent review (Förster and Woods, 2012).

4. The steroid-ETH-juvenile hormone triad in reproduction

4.1 ETH functions as an allatotropin

While ETHs are named for obligatory functions served during developmental transitions, the persistence of Inka cells through metamorphosis to the adult stage was observed years ago and has been a subject of curiosity since that time (Park et al., 2002). An early report showed expression of ETH receptor transcripts in corpora allata of the silkworm, suggesting an allatoregulatory role for ETH (Yamanaka et al., 2008). More recent reports have implicated ETHs as allatotropins in the adult stage with important functions in coupling ecdysone signalling with juvenile hormones important in reproductive physiology and associated behaviours. Using the Aug21-Gal4 driver known to target corpora allata (Riddiford et al., 2010), Deshpande produced JH deficiency selectively in male *Drosophila* following RNAi knockdown of ETHR (Deshpande, 2012). Further evidence supporting an allatoregulatory role for ETH came from studies in the yellow fever mosquito, *Aedes aegypti*, where ETHR transcripts were detected in the CA and evidence was presented showing that the peptide promotes JH production through calcium-mediated augmentation of the enzyme juvenile hormone acid methyltransferase (JHAMT) (Areiza et al., 2014). The mechanism underlying the allatotropic function was demonstrated as release of calcium from IP_3 sensitive stores in CA cells. It was concluded that ETH release at eclosion ensures proper timing of JH release in pharate adult mosquitoes. However, the question of posteclosion, adult related physiological roles for ETH remained a question that was addressed in a series of subsequent studies.

In *Drosophila* adults, Inka cells occur as bilateral pairs, two in the anterior thorax in close proximity to the CA and 7 pairs in the dorsal abdomen (Fig. 2) (Lee et al., 2017; Meiselman et al., 2017). In females, Inka cells are particularly clustered proximal to the ovaries. ETH along with ETHR-A and ETHR-B receptor isoforms are expressed throughout the first 3 weeks of adulthood as measured by RT-PCR. In females, ETHR-B expression predominates, whereas the two receptor isoforms are more or less equally expressed in males. The CA respond to ETH exposure with robust calcium mobilization, which is diminished upon ETHR knockdown. Pronounced JH deficiency caused by ETHR silencing results in a ~30% drop in egg production similar to the reduction caused by direct ablation of the CA. Reduction of fecundity was rescued by methoprene treatment. As demonstrated previously during moulting of juvenile insects, steroids regulate ETH and ETHR expression in the adult stage. Knockdown of EcR specifically in the Inka cell results in significant reduction of egg production, although not to the extent observed following ETHR knockdown in the CA. This reduction is interpreted to occur through reduced synthesis of

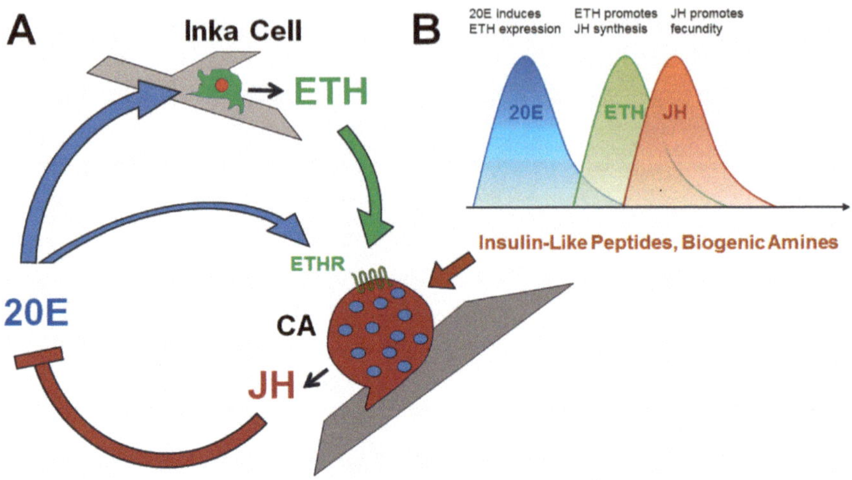

Fig. 5 The ecdysone (20E)-ETH-JH signalling triad in adult *Drosophila* (A) 20E induces expression of ETH in Inka cells and ETHR in target tissues, including the corpora allata (CA). The CA integrates ETH and other signals to determine levels of JH. (B) 20E induces ETH expression in the Inka cell, but peptide release occurs only after steroid levels fall. ETH release promotes JH and normal levels of fecundity. *From Meiselman, M., Lee, S.-S., Tran, R.-T., Dai, H., Ding, Y., Rivera-Perez, C., Wijesekera, T.P., Dauwalder, B., Noriega, F.G., Adams, M.E., 2017. Endocrine network essential for reproductive success in Drosophila melanogaster. Proc. Natl. Acad. Sci. U. S. A. 114, E3849–E3858. https://doi.org/10.1073/pnas.1620760114.*

ETH in the Inka cell. Impaired ETH signalling also reduced reproductive potential in male flies. For these experiments, ETH/JH deficient males were mated with wild-type females; such matings also resulted in lower fecundity. Factors responsible for this reduction in male potential remain to be investigated. These findings point to a signalling cascade involving steroid promotion of ETH expression, which in turn serves to augment juvenile hormone signalling contributing to normal levels of fecundity (Fig. 5).

The roles of ETH and JH in reproduction also have been examined in the oriental fruit fly, *Bactrocera dorsalis* (Shi et al., 2019). In this study employing quantitative RT-qPCR, expression of ETHR-B was significant, peaking at 10–12 days posteclosion, whereas ETHR-A was not detected in significant quantities. Again, impaired ETH signalling resulted in lower fecundity. This report emphasizes the importance of the ETHR-B isoform in adult reproductive performance.

4.2 A steroid-ETH-JH triad in regulation of reproductive responses to stress

In view of the findings that ETH functions as an allatotropin influencing reproduction in flies and mosquitoes, it was of interest to ask how stress-induced elevation of ecdysteroid levels might affect ETH and JH levels. From numerous studies of ecdysis over the years, it is known that elevated steroid levels drive ETH and ETHR expression, but block ETH release by Inka cells. If the same relationship between steroid and ETH secretion holds during the adult stage, one might hypothesize that stress-induced steroid elevation would create ETH and hence JH deficiencies, thereby contributing to lower fecundity under stressful conditions as observed previously (Gruntenko et al., 2003; Terashima and Bownes, 2004; Terashima et al., 2005).

This hypothesis was tested by imposing heat and nutritional stresses on adult female flies. These conditions indeed elevated steroid levels and created ETH deficiency as measured by enzyme immunoassay (Meiselman et al., 2018). Phenotypes included arrested oogenesis, egg retention in ovaries, suppression of ovulation, and reduced fecundity. Using several approaches to restore ETH levels rescued these deficiency phenotypes. These included direct injections of ETH, Trp-A1-mediated release of ETH from Inka cells, knockdown or inhibition of Inka cell EcR expression, and overexpression of βFTZ-F1 in Inka cells to restore secretory competence. Interestingly, this study revealed for the first time that ETH activates octopamine neurons innervating oviducts and that silencing of ETHR expression in these neurons attenuates these responses. It was also noteworthy that elevated steroid levels, in this instance induced by stress, appears to inhibit secretory competence of adult stage Inka cells, creating ETH and JH deficiencies. The finding that overexpression of βFTZ-F1 in Inka cells overcomes reproductive deficits provides support for this conclusion (Meiselman et al., 2018).

Taken together, these studies demonstrate that ETH signalling plays important functional roles in *Drosophila* reproduction not only by promoting JH production, but through direct actions on the reproductive tract to promote ovulation (Meiselman et al., 2017, 2018). Furthermore, ETH appears to provide a link between steroid elevation and re-allocation of energy from reproductive purposes to survival under stressful conditions, both with respect to 1. ETH as an allatotropin, which under stress translates to ETH and hence JH deficiency, and 2. Direct actions of the peptide on the reproductive tract to promote egg production, which under stress-induced steroid elevation and resulting ETH deficiency translates into reduced ovulation and egg production.

5. ETH signalling in adult male reproductive behaviours

5.1 ETHR regulation of olfactory interneurons

Knockdown of ETHR in the CA using the Aug21-Gal4 driver leads to male-specific JH deficiency and pronounced elevation of male-male courtship behaviour (Deshpande et al., 2018). No evidence could be found to relate JH deficiency to this courtship behaviour phenotype; in particular, methoprene treatment has no effect on this behavioural outcome. Moreover, use of a more specific CA-directed Gal4 line—JHAMT-Gal4 for ETHR knockdown, known to cause JH deficiency (Lee et al., 2017; Meiselman et al., 2017), fails to produce the male-male courtship phenotype. It soon became clear that the Aug21-Gal4 driver, reputed to be specific CA driver in pre-metamorphic stages of *Drosophila*, labels extensively in the adult CNS. Indeed, some of these elements proved to be fruitless-positive and likely to account for the courtship phenotype.

Among the CNS neuronal elements strongly labelled by Aug21-Gal4 are antennal interneurons and a subset of these express ETHR. Use of several Gal4 lines that label antennal lobe interneurons more specifically to knock down ETHR expression caused elevated male-male courtship behaviours. In particular, the NP3056-Gal4 driver, which is highly specific for antennal lobe interneurons (Chou et al., 2010), re-capitulated the Aug21-Gal4 male-male courtship phenotype (Deshpande et al., 2018). These findings indicate that ETHR signalling in the olfactory system during adult male life is important in regulation of male courtship behaviour. This study was one of the first to confirm the functional significance of ETH signalling in the brain of adult *Drosophila*.

5.2 An ETH-JH-dopamine cascade is essential for short-term courtship memory

Exposure of *Drosophila* males to rejection by mated females leads to altered courtship behaviour in future encounters with receptive virgin females. Such rejected males exhibit a "courtship memory" proportional to duration of such rejection, manifested as reduced motivation for courtship (Siegel and Hall, 1979). Such prior experience can be viewed as a type of social stress, which has been tied to elevated ecdysteroid levels and 20E signalling via the non-canonical GPCR-based DopEcR subtype mediating cAMP signalling (Ishimoto et al., 2009, 2013).

Recent studies show that the epitracheal endocrine system and JH signalling also contribute significantly to hormonal state-dependent courtship memory. The first of these studies by Sang Soo Lee focused on short-term courtship memory (STM), whereby males exposed to 1 h of rejection by a mated female exhibited reduced courtship motivation for several hours (Lee et al., 2017). Courtship STM is abolished following Inka cell ablation or block of ETH secretion via expression of tetanus toxin. Similarly, RNAi knockdown of ETH receptors in the CA causes loss of STM through JH deficiency. Methoprene treatment rescues of ETH/JH deficiency-induced loss of STM. This evidence indicates that ETH-dependent JH signalling is required for STM.

Evidence also indicates that the memory defect involves shortened memory retention rather than acquisition and is attributable to JH-mediated activity in tyrosine hydroxylase (TH)-expressing dopamine neurons in the brain. ETH/JH-dependent STM occurs during a critical period 2–3 days post-eclosion. Furthermore, STM appears to be predominantly based on female rejection behaviour rather than olfactory cues displayed by mated females such as the CVA. Thus, a cascade ETH-JH-dopamine signalling operating during an early adult critical period is required for social context-dependent memory performance.

5.3 ETH and JH are required for long-term courtship memory

In a subsequent study, the influence of hormonal state on long-term courtship memory (LTM) was assessed. In this case, males conditioned to 5 h of rejection by mated females develop courtship memory lasting in excess of 24 h. Evidence from this study indicates that, in contrast to its role in STM, ETH exerts direct influences on brain neurons as well as indirectly via JH signalling. Neither of these signalling cascades involves TH-expressing dopamine neurons (Lee and Adams, 2021).

As indicated for STM, ETH availability is required for LTM, since either Inka cell ablation or block of secretion by conditional, temperature-sensitive *shibire* expression abolishes this form of courtship memory. In addition, ETHR knockdown specifically in mushroom body (MB) γ-lobe neurons abolishes LTM. Likewise, ETHR knockdown in octopaminergic dorsal-anterior-lateral (DAL) neurons, previously implicated in associative LTM (Chen et al., 2012) was sufficient to prevent LTM. Exposure of the fly brain to ETH results in calcium mobilization in the MB γ-lobe as well as DAL neurons. Likewise, knockdown of the JH receptor Met (but not gce) in mushroom body γ-lobe neurons blocks LTM. These data indicate that ETH and JH play dual roles by converging in the MB to facilitate LTM.

In the study of LTM by Lee and Adams (2021), the minimum continuous training period required for LTM induction was found to be 5 h. They found, however, that augmentation of ETH levels via Trp-A1 activation of Inka cells reduces this minimum training period significantly (Lee and Adams, 2021), indicating that ETH is not only required for LTM, but facilitates it by reducing the training period required.

6. Conclusions

Since the discovery of Inka cells as the source of ETH, the epitracheal endocrine system has proved to be of central importance in events that terminate moulting with initiation and scheduling of the ecdysis sequence. In order to advance successfully to the next stage, successive waves of gene expression prepare Inka cells for ETH release and cellular targets for responsiveness to this hormone. This occurs through early steroid-induced expression of genes encoding ETH and ETHR and late expression of βftz-f1 required for Inka cell secretory competence. Timing of ETH release must be delayed as a protective measure until the animal is developmentally and physiologically "ready-to-go." Once ETH is released, it acts in concert with EH, kinin, CCAP, bursicon, and a plethora of modulators involving feedback and feedforward signalling to facilitate tracheal airfilling and shedding of the old cuticle. The pronounced stereotypy of the ecdysis sequence makes it a favourable paradigm for analysis of how genes construct and schedule innate behaviours. In addition, the modulatory nature of ecdysis control provides opportunities for gaining insights into the phylogenetic evolution of innate behaviours.

Inka cells are now known to be re-purposed for endocrine functions during adulthood, including promotion of reproductive physiology, responses to stress, and behaviours related to olfaction, sex recognition, and courtship memory.

Acknowledgement

Research in the author's laboratory cited in this work was supported by funding provided by the National Institutes of Health R01 GM-067310.

References

Adams, M.E., Zitnan, D., 1997. Identification of ecdysis-triggering hormone in the silkworm Bombyx mori. Biochem. Biophys. Res. Commun. 230, 188–191. https://doi.org/10.1006/bbrc.1996.5915.

Akai, H., 1992. Ultrastructure of Epitracheal gland during larval-pupal molt of Bombyx mori. Cytologia 57, 195–201. https://doi.org/10.1508/cytologia.57.195.

Areiza, M., Nouzova, M., Rivera-Perez, C., Noriega, F.G., 2014. Ecdysis triggering hormone ensures proper timing ofjuvenile hormone biosynthesis in pharate adult mosquitoes. Insect Biochem. Mol. Biol. 54, 98–105. https://doi.org/10.1016/j.ibmb.2014.09.006.

Assmann, M., Kuhn, A., Dürrnagel, S., Holstein, T.W., Gründer, S., 2014. The comprehensive analysis of DEG/ENaC subunits in Hydrareveals a large variety of peptide-gated channels, potentially involved in neuromuscular transmission. BMC Biol. 12, 84. https://doi.org/10.1186/s12915-014-0084-2.

Baker, J.D., McNabb, S.L., Truman, J.W., 1999. The hormonal coordination of behavior and physiology at adult ecdysis in Drosophila melanogaster. J. Exp. Biol. 202, 3037–3048.

Chang, J.-C., Yang, R.-B., Adams, M.E., Lu, K.-H., 2009. Receptor guanylyl cyclases in Inka cells targeted by eclosion hormone. Proc. Natl. Acad. Sci. U. S. A. 106, 13371–13376. https://doi.org/10.1073/pnas.0812593106.

Chen, C.-C., Wu, J.-K., Lin, H.-W., Pai, T.-P., Fu, T.-F., Wu, C.-L., Tully, T., Chiang, A.-S., 2012. Visualizing long-term memory formation in two neurons of the Drosophila brain. Science (New York, N.Y.) 335, 678–685. https://doi.org/10.1126/science.1212735.

Cho, K.-H., Daubnerová, I., Park, Y., Zitnan, D., Adams, M.E., 2013. Secretory competence in a gateway endocrine cell conferred by the nuclear receptor βFTZ-F1 enables stage-specific ecdysone responses throughout development in Drosophila. Dev. Biol. 385, 253–262. https://doi.org/10.1016/j.ydbio.2013.11.003.

Chou, Y.-H., Zheng, X., Beachy, P.A., Luo, L., 2010. Patterning axon targeting of olfactory receptor neurons by coupled hedgehog signaling at two distinct steps. Cell 142, 954–966. https://doi.org/10.1016/j.cell.2010.08.015.

Christie, A.E., Chi, M., 2015. Identification of the first neuropeptides from the enigmatic hexapod order Protura. Gen. Comp. Endocrinol. 224, 18–37. https://doi.org/10.1016/j.ygcen.2015.05.015.

Clark, A.C., del Campo, M.L., Ewer, J., 2004. Neuroendocrine control of larval ecdysis behavior in Drosophila: complex regulation by partially redundant neuropeptides. J. Neurosci. 24, 4283–4292. https://doi.org/10.1523/jneurosci.4938-03.2004.

Dai, H., Ma, L., Wang, J., Jiang, R., Wang, Z., Fei, J., 2008. Knockdown of ecdysis-triggering hormone gene with a binary UAS/GAL4 RNA interference system leads to lethal ecdysis deficiency in silkworm. Acta Biochim. Biophys. Sin. 40, 790–795.

Daubnerova, I., Roller, V., Satake, H., Zhang, C., Kim, Y.-J., Zitnan, D., 2021. Identification and function of ETH receptor networks in the silkworm, *Bombyx mori*. Sci. Rep. 11, 11693–11716. https://doi.org/10.1038/s41598-021-91022-8.

Deshpande, S.A., 2012. Ecdysis Triggering Hormone Signaling in *Drosophila melanogaster*. Ph.D Dissertation, University of California, Riverside.

Deshpande, S.A., Meiselman, M., Hice, R.H., Arensburger, P., Rivera-Perez, C., Kim, D.-H., Croft, R.L., Noriega, F.G., Adams, M.E., 2018. Ecdysis triggering hormone receptors regulate male courtship behavior via antennal lobe interneurons in Drosophila. Gen. Comp. Endocrinol. 278, 79–88. https://doi.org/10.1016/j.ygcen.2018.12.003.

Diao, F., Mena, W., Shi, J., Park, D., Diao, F., Taghert, P., Ewer, J., White, B.H., 2015a. The splice isoforms of the drosophila ecdysis triggering hormone receptor have developmentally distinct roles. Genetics 202, 175–189. https://doi.org/10.1534/genetics.115.182121.

Diao, F., Ironfield, H., Luan, H., Diao, F., Shropshire, W.C., Ewer, J., Marr, E., Potter, C.J., Landgraf, M., White, B.H., 2015b. Plug-and-play genetic access to drosophila cell types using exchangeable exon cassettes. Cell Rep. 10, 1410–1421. https://doi.org/10.1016/j.celrep.2015.01.059.

Diao, F., Elliott, A.D., Diao, F., Shah, S., White, B.H., 2017. Neuromodulatory connectivity defines the structure of a behavioral neural network. Elife 6, e29797. https://doi.org/10.7554/eLife.29797.020.

Ewer, J., Gammie, S.C., Truman, J.W., 1997. Control of insect ecdysis by a positive-feedback endocrine system: roles of eclosion hormone and ecdysis triggering hormone. J. Exp. Biol. 200, 869–881.

Förster, T.D., Woods, H.A., 2012. Mechanisms of tracheal filling in insects. Biol. Rev. Camb. Philos. Soc. 88, 1–14. https://doi.org/10.1111/j.1469-185x.2012.00233.x.

Fuse, M., Truman, J.W., 2002. Modulation of ecdysis in the moth Manduca sexta: the roles of the suboesophageal and thoracic ganglia. J. Exp. Biol. 205, 1047–1058.

Gammie, S.C., Truman, J.W., 1997. Neuropeptide hierarchies and the activation of sequential motor behaviors in the hawkmoth, Manduca sexta. J. Neurosci. 17, 4389–4397.

Gammie, S.C., Truman, J.W., 1999. Eclosion hormone provides a link between ecdysis-triggering hormone and crustacean cardioactive peptide in the neuroendocrine cascade that controls ecdysis behavior. J. Exp. Biol. 202, 343–352.

Gauthier, S.A., VanHaaften, E., Cherbas, L., Cherbas, P., Hewes, R.S., 2012. Cryptocephal, the Drosophila melanogaster ATF4, is a specific coactivator for ecdysone receptor isoform B2. PLoS Genet. 8, e1002883. https://doi.org/10.1371/journal.pgen.1002883.

Gruntenko, N.E., Bownes, M., Terashima, J., Sukhanova, M.Z., Raushenbach, I.Y., 2003. Heat stress affects oogenesis differently in wild-type Drosophila virilis and a mutant with altered juvenile hormone and 20-hydroxyecdysone levels. Insect Mol. Biol. 12, 393–404.

Gruntenko, N.E., Wen, D., Karpova, E.K., Adonyeva, N.V., Liu, Y., He, Q., Faddeeva, N.V., Fomin, A.S., Li, S., Rauschenbach, I.Y., 2010. Altered juvenile hormone metabolism, reproduction and stress response in Drosophila adults with genetic ablation of the corpus allatum cells. Insect Biochem. Mol. Biol. 40, 891–897. https://doi.org/10.1016/j.ibmb.2010.09.001.

Ikeda, E., 1913. Kimon Rimensen. In: Ikeda, E. (Ed.), Experimental Anatomy and Physiology of Bombyx mori. Meibundo, Tokyo, pp. 242–243.

Ishimoto, H., Sakai, T., Kitamoto, T., 2009. Ecdysone signaling regulates the formation of long-term courtship memory in adult Drosophila melanogaster. Proc. Natl. Acad. Sci. U. S. A. 106, 6381–6386. https://doi.org/10.1073/pnas.0810213106.

Ishimoto, H., Wang, Z., Rao, Y., Wu, C.-F., Kitamoto, T., 2013. A novel role for ecdysone in Drosophila conditioned behavior: linking GPCR-mediated non-canonical steroid action to cAMP signaling in the adult brain. PLoS Genet. 9, e1003843. https://doi.org/10.1371/journal.pgen.1003843.

Iversen, A., Cazzamali, G., Williamson, M., Hauser, F., Grimmelikhuijzen, C.J.P., 2002. Molecular identification of the first insect ecdysis triggering hormone receptors. Biochem. Biophys. Res. Commun. 299, 924–931. https://doi.org/10.1016/s0006-291x(02)02798-5.

Kataoka, H., Troetschler, R.G., Kramer, S.J., Cesarin, B.J., Schooley, D.A., 1987. Isolation and primary structure of the eclosion hormone of the tobacco hornworm, Manduca sexta. Biochem. Biophys. Res. Commun. 146, 746–750. https://doi.org/10.1016/0006-291x(87)90592-4.

Kim, Y.-J., Spalovská-Valachová, I., Cho, K.-H., Zitnanová, I., Park, Y., Adams, M.E., Zitnan, D., 2004. Corazonin receptor signaling in ecdysis initiation. Proc. Natl. Acad. Sci. U. S. A. 101, 6704–6709. https://doi.org/10.1073/pnas.0305291101.

Kim, Y.-J., Zitnan, D., Cho, K.-H., Schooley, D.A., Mizoguchi, A., Adams, M.E., 2006a. Central peptidergic ensembles associated with organization of an innate behavior. Proc. Natl. Acad. Sci. U. S. A. 103, 14211–14216. https://doi.org/10.1073/pnas.0603459103.

Kim, Y.-J., Zitnan, D., Galizia, C.G., Cho, K.-H., Adams, M.E., 2006b. A command chemical triggers an innate behavior by sequential activation of multiple peptidergic ensembles. Curr. Biol. 16, 1395–1407. https://doi.org/10.1016/j.cub.2006.06.027.

Kim, D.-H., Han, M.-R., Lee, G., Lee, S.-S., Kim, Y.-J., Adams, M.E., 2015. Rescheduling behavioral subunits of a fixed action pattern by genetic manipulation of Peptidergic signaling. PLoS Genet. 11, e1005513. https://doi.org/10.1371/journal.pgen.1005513.

Kim, D.-H., Kim, Y.-J., Adams, M.E., 2018. Endocrine regulation of airway clearance in Drosophila. Proc. Natl. Acad. Sci. U. S. A. 129, 201717257. https://doi.org/10.1073/pnas.1717257115.

Kingan, T.G., Adams, M.E., 2000. Ecdysteroids regulate secretory competence in Inka cells. J. Exp. Biol. 203, 3011–3018.

Kingan, T.G., Gray, W., Zitnan, D., Adams, M.E., 1997. Regulation of ecdysis-triggering hormone release by eclosion hormone. J. Exp. Biol. 200, 3245–3256.

Klein, C., Kallenborn, H.G., Radlicki, C., 1999. The 'Inka cell' and its associated cells: ultrastructure of the epitracheal glands in the gypsy moth, Lymantria dispar. J. Insect Physiol. 45, 65–73. https://doi.org/10.1016/s0022-1910(98)00085-7.

Krüger, E., Mena, W., Lahr, E.C., Johnson, E.C., Ewer, J., 2015. Genetic analysis of Eclosion hormone action during Drosophila larval ecdysis. Development 142, 4279–4287. https://doi.org/10.1242/dev.126995.

Lahr, E.C., Dean, D., Ewer, J., 2012. Genetic analysis of ecdysis behavior in Drosophila reveals partially overlapping functions of two unrelated neuropeptides. J. Neurosci. 32, 6819–6829. https://doi.org/10.1523/jneurosci.5301-11.2012.

Lee, S.S., Adams, M.E., 2021. Regulation of Drosophila long-term courtship memory by Ecdysis triggering hormone. Front. Neurosci. 15, 670322. https://doi.org/10.3389/fnins.2021.670322.

Lee, S.-S., Ding, Y., Karapetians, N., Rivera-Perez, C., Noriega, F.G., Adams, M.E., 2017. Hormonal signaling Cascade during an early-adult critical period required for courtship memory retention in Drosophila. Curr. Biol. 27, 2798–2809.e3. https://doi.org/10.1016/j.cub.2017.08.017.

Lenaerts, C., Cools, D., Verdonck, R., Verbakel, L., Broeck, J.V., Marchal, E., 2017. The ecdysis triggering hormone system is essential for successful moulting of a major hemimetabolous pest insect, Schistocerca gregaria. Sci. Rep. 7, 46502. https://doi.org/10.1038/srep46502.

Marti, T., Takio, K., Walsh, K.A., Terzi, G., Truman, J.W., 1987. Microanalysis of the amino acid sequence of the eclosion hormone from the tobacco hornworm Manduca sexta. FEBS Lett. 219, 415–418. https://doi.org/10.1016/0014-5793(87)80263-6.

McNabb, S.L., Baker, J.D., Agapite, J., Steller, H., Riddiford, L.M., Truman, J.W., 1997. Disruption of a behavioral sequence by targeted death of peptidergic neurons in Drosophila. Neuron 19, 813–823.

Meiselman, M., Lee, S.-S., Tran, R.-T., Dai, H., Ding, Y., Rivera-Perez, C., Wijesekera, T.P., Dauwalder, B., Noriega, F.G., Adams, M.E., 2017. Endocrine network essential for reproductive success in Drosophila melanogaster. Proc. Natl. Acad. Sci. U. S. A. 114, E3849–E3858. https://doi.org/10.1073/pnas.1620760114.

Meiselman, M.R., Kingan, T.G., Adams, M.E., 2018. Stress-induced reproductive arrest in Drosophila occurs through ETH deficiency-mediated suppression of oogenesis and ovulation. BMC Biol. 16, 239. https://doi.org/10.1186/s12915-018-0484-9.

Mena, W., Diegelmann, S., Wegener, C., Ewer, J., 2016. Stereotyped responses of Drosophila peptidergic neuronal ensemble depend on downstream neuromodulators. Elife 5, e19686. https://doi.org/10.7554/elife.19686.

Nässel, D.R., Zandawala, M., 2019. Recent advances in neuropeptide signaling in Drosophila, from genes to physiology and behavior. Prog. Neurobiol. 179, 101607. https://doi.org/10.1016/j.pneurobio.2019.02.003.

Nässel, D.R., Zandawala, M., 2020. Hormonal axes in Drosophila: regulation of hormone release and multiplicity of actions. Cell Tissue Res. 382, 233–266. https://doi.org/10.1007/s00441-020-03264-z.

Park, Y., Zitnan, D., Gill, S.S., Adams, M.E., 1999. Molecular cloning and biological activity of ecdysis-triggering hormones in Drosophila melanogaster. FEBS Lett. 463, 133–138.

Park, Y., Filippov, V., Gill, S.S., Adams, M.E., 2002. Deletion of the ecdysis-triggering hormone gene leads to lethal ecdysis deficiency. Development 129, 493–503.

Park, Y., Kim, Y.-J., Dupriez, V., Adams, M.E., 2003. Two subtypes of ecdysis-triggering hormone receptor in Drosophila melanogaster. J. Biol. Chem. 278, 17710–17715. https://doi.org/10.1074/jbc.m301119200.

Riddiford, L.M., Truman, J.W., Mirth, C.K., Shen, Y.-C., 2010. A role for juvenile hormone in the prepupal development of Drosophila melanogaster. Development 137, 1117–1126. https://doi.org/10.1242/dev.037218.

Roller, L., Zitnanová, I., Dai, L., Simo, L., Park, Y., Satake, H., Tanaka, Y., Adams, M.E., Zitnan, D., 2010. Ecdysis triggering hormone signaling in arthropods. Peptides 31, 429–441. https://doi.org/10.1016/j.peptides.2009.11.022.

Schleidt, W.M., 1974. How "fixed" is the fixed action pattern? Z. Tierpsychol. 36, 184–211.

Scott, R.L., Diao, F., Silva, V., Park, S., Luan, H., Ewer, J., White, B.H., 2020. Non-canonical Eclosion hormone-expressing cells regulate Drosophila Ecdysis. Iscience 23, 101108. https://doi.org/10.1016/j.isci.2020.101108.

Shen, C.-H., Xu, Q.-Y., Fu, K.-Y., Guo, W.-C., Jin, L., Li, G.-Q., 2020. Two splice isoforms of Leptinotarsa Ecdysis triggering hormone receptor have distinct roles in larva-pupa transition. Front. Physiol. 11, 593962. https://doi.org/10.3389/fphys.2020.593962.

Shen, C.-H., Xu, Q.-Y., Fu, K.-Y., Guo, W.-C., Jin, L., Li, G.-Q., 2021. Ecdysis triggering hormone is essential for larva–pupa–adult transformation in Leptinotarsa decemlineata. Insect Mol. Biol. 30, 241–252. https://doi.org/10.1111/imb.12691.

Shi, Y., Jiang, H.-B., Gui, S.-H., Liu, X.-Q., Pei, Y.-X., Xu, L., Smagghe, G., Wang, J.-J., 2017. Ecdysis triggering hormone signaling (ETH/ETHR-A) is required for the larva-larva Ecdysis in Bactrocera dorsalis (Diptera: Tephritidae). Front. Physiol. 8, 984. https://doi.org/10.3389/fphys.2017.00587.

Shi, Y., Liu, T.-Y., Jiang, H.-B., Liu, X.-Q., Dou, W., Park, Y., Smagghe, G., Wang, J.-J., 2019. The Ecdysis triggering hormone system, via ETH/ETHR-B, is essential for successful reproduction of a major pest insect, Bactrocera dorsalis (Hendel). Front. Physiol. 10, 151. https://doi.org/10.3389/fphys.2019.00151.

Siegel, R.W., Hall, J.C., 1979. Conditioned responses in courtship behavior of normal and mutant Drosophila. Proc. Natl. Acad. Sci. U. S. A. 76, 3430–3434. https://doi.org/10.1073/pnas.76.7.3430.

Taghert, P.H., Nitabach, M.N., 2012. Peptide neuromodulation in invertebrate model systems. Neuron 76, 82–97. https://doi.org/10.1016/j.neuron.2012.08.035.

Tanaka, S., 2006. Corazonin and locust phase polyphenism. Appl. Entomol. Zool. 41, 179–193. https://doi.org/10.1303/aez.2006.179.

Terashima, J., Bownes, M., 2004. Translating available food into the number of eggs laid by Drosophila melanogaster. Genetics 167, 1711–1719. https://doi.org/10.1534/genetics.103.024323.

Terashima, J., Takaki, K., Sakurai, S., Bownes, M., 2005. Nutritional status affects 20-hydroxyecdysone concentration and progression of oogenesis in Drosophila melanogaster. J. Endocrinol. 187, 69–79. https://doi.org/10.1677/joe.1.06220.

Truman, J.W., 1992a. Chapter 30 The eclosion hormone system of insects. In: The Peptidergic Neuron. Elsevier, pp. 361–374.

Truman, J.W., 1992b. The eclosion hormone system of insects. Prog. Brain Res. 92, 361–374. https://doi.org/10.1016/s0079-6123(08)61189-9.

White, B.H., Ewer, J., 2014. Neural and hormonal control of postecdysial behaviors in insects. Annu. Rev. Entomol. 59, 363–381. https://doi.org/10.1146/annurev-ento-011613-162028.

Yamada, M., Murata, T., Hirose, S., Lavorgna, G., Suzuki, E., Ueda, H., 2000. Temporally restricted expression of transcription factor betaFTZ-F1: significance for embryogenesis, molting and metamorphosis in Drosophila melanogaster. Development 127, 5083–5092.

Yamanaka, N., Yamamoto, S., Zitnan, D., Watanabe, K., Kawada, T., Satake, H., Kaneko, Y., Hiruma, K., Tanaka, Y., Shinoda, T., Kataoka, H., 2008. Neuropeptide receptor transcriptome reveals unidentified neuroendocrine pathways. PLoS One 3, e3048. https://doi.org/10.1371/journal.pone.0003048.

Yao, Z., Macara, A.M., Lelito, K.R., Minosyan, T.Y., Shafer, O.T., 2012. Analysis of functional neuronal connectivity in the Drosophila brain. J. Neurophysiol. 108, 684–696. https://doi.org/10.1152/jn.00110.2012.

Zhu, L., Zhang, W., Li, G., Sun, Q.-Z., Wang, J.-J., Smagghe, G., Jiang, H.-B., 2019. Molecular characterization of ecdysis triggering hormone and its receptor in citrus red mite (Panonychus citri). Comp. Biochem. Physiol. A Mol. Integr. Physiol. 230, 100–105. https://doi.org/10.1016/j.cbpa.2019.01.003.

Zitnan, D., 1989. Regulatory Peptides and Peptidergic Organs of Insects. Ph.D dissertation, Slovak Academy of Sciences, Bratislava.

Zitnan, D., Adams, M.E., 2000. Excitatory and inhibitory roles of central ganglia in initiation of the insect ecdysis behavioural sequence. J. Exp. Biol. 203, 1329–1340.

Zitnan, D., Adams, M.E., 2012. Neuroendocrine regulation of Ecdysis. In: Gilbert, L.I. (Ed.), Insect Endocrinology. Elsevier, pp. 253–309.

Zitnan, D., Kingan, T.G., Hermesman, J.L., Adams, M.E., 1996. Identification of ecdysis-triggering hormone from an epitracheal endocrine system. Science (New York, N.Y.) 271, 88–91. https://doi.org/10.1126/science.271.5245.88.

Zitnan, D., Ross, L.S., Zitnanová, I., Hermesman, J.L., Gill, S.S., Adams, M.E., 1999. Steroid induction of a peptide hormone gene leads to orchestration of a defined behavioral sequence. Neuron 23, 523–535.

Zitnan, D., Hollar, L., Spalovská, I., Takác, P., Zitnanová, I., Gill, S.S., Adams, M.E., 2002. Molecular cloning and function of ecdysis-triggering hormones in the silkworm. J. Exp. Biol. 205, 1–15.

Zitnan, D., Zitnanová, I., Spalovská, I., Takác, P., Park, Y., Adams, M.E., 2003. Conservation of ecdysis-triggering hormone signalling in insects. J. Exp. Biol. 206, 1275–1289. https://doi.org/10.1242/jeb.00261.

Zitnanová, I., Adams, M.E., Zitnan, D., 2001. Dual ecdysteroid action on the epitracheal glands and central nervous system preceding ecdysis of Manduca sexta. J. Exp. Biol. 204, 3483–3495.

CHAPTER FOUR

Regulation of sleep in *Drosophila melanogaster*

Juan Du, Pengfei Lv, Tiantian Fu, Yu Wei, Yahong Li, Zhe Liu, Lei He, Qiankun He, and Zhangwu Zhao*
Department of Entomology and MOA Key Lab of Pest Monitoring and Green Management, College of Plant Protection, China Agricultural University, Beijing, China
*Corresponding author: e-mail address: zhaozw@cau.edu.cn

Contents

Advances in Insect Physiology, Volume 60
ISSN 0065-2806
https://doi.org/10.1016/bs.aiip.2021.04.001

Abstract

Studies on mechanisms of sleep regulation in the fruit fly, *Drosophila melanogaster* have moved the field forward considerably during the past 20 years. In this review, important advances in our understanding of sleep regulation are summarized according to types of molecules involved. Such molecules mediate influences of photoperiod, temperature, nutrition, and different types of disease. These advances on the genetic control of sleep in *Drosophila* provide crucial insights into the mechanisms of sleep regulation.

1. Introduction

Studies on insect sleep have focused largely on *Drosophila melanogaster*, which as a model genetic organism is amenable to manipulation of a wide range of biological processes, from physiological development to behaviour. Discovery of roles played by the *period* gene in rhythmic control of locomotory behaviour and eclosion ushered in a new era of genetic study of circadian rhythm in *Drosophila* in 1970s (Konopka and Benzer, 1971). However, genetic studies on sleep have been around for only the last 20 years (Hendricks et al., 2000; Shaw et al., 2000). Although measures of non-rapid eye movement (NREM) and rapid eye movement (REM) relevant to human sleep are not possible in flies, local field potential (LFP) recordings show that *Drosophila* sleep includes at least two distinct stages: One is increased oscillatory activity during sleep induction, while subsequent stages involve desynchronized or decreased brain activity (Yap et al., 2017).

Sleep is regulated by both circadian rhythm and homeostatic factors. They are controlled by the core oscillator in brain clock neurons and the central complex (ellipsoid body and fan-shaped body) respectively, which receives and synthesizes signals from ambient environmental factors such as photoperiod, temperature, nutrition, and so on. In the *Drosophila* brain, 150 clock neurons are devoted to circadian rhythm control, while sleep homeostasis is regulated by the mushroom body and central complex. Circadian rhythm and homeostasis interact to modulate sleep, in which most neural circuits and pathways connecting clock neurons and the central complex have been clarified (Guo et al., 2016, 2018; Potdar and Sheeba, 2018).

In this review, we summarize advances in sleep regulation, in which involved genes are classified according to their function (Table 1). In addition, functional molecules involved in different sleep regulation-associated

Table 1 Summary of sleep regulators in *Drosophila*.

Type	Pathway/molecule	Effect on sleep	Functional tissues	References
Circadian regulating genes	Pigment-dispersing factor (PDF)	Wake promoting	Clock neurons	Parisky et al. (2008)
	Period (Per)	Inhibiting sleep homeostasis	Clock neurons	Shaw et al. (2002)
	Clock (Clk)	Inhibiting sleep homeostasis	Clock neurons	Shaw et al. (2002)
	Cycle (Cyc)	Inhibiting sleep homeostasis	Clock neurons	Shaw et al. (2002)
	Timeless (Tim)	Inhibiting sleep homeostasis	Clock neurons	Hendricks et al. (2000), Shaw et al. (2002)
	Resistant to dieldrin (Rdl)	Sleep promoting	Clock neurons	Parisky et al. (2008)
	Shaker cognate 1 (Shal/Kv4)	Promoting sleep onset at dusk	Clock neurons	Feng et al. (2018)
	Wide awake	Promoting sleep onset at dusk	Clock neurons	Liu et al. (2014)
	Diuretic hormone 31 (DH31)	Wake promoting (Night)	Clock neurons	Kunst et al. (2014)
	AstA	Sleep promoting	Clock neurons	Donlea et al. (2018), Ni et al. (2019)

Continued

Table 1 Summary of sleep regulators in *Drosophila*.—cont'd

Type	Pathway/molecule	Effect on sleep	Functional tissues	References
Function of neurotransmitters in promoting wake	Dopamine	Wake promoting	Mushroom bodies (MB) and the fan-shaped body (FB), ellipsoid body (EB), noduli (NO), protocerebral bridge (PB)	Wisor et al. (2001), Rye and Jankovic (2002), Friggi-Grelin et al. (2003), Isaac and Berridge (2003), Andretic et al. (2005), Kume et al. (2005), Pitman et al. (2006), Andretic et al. (2008), Mao and Davis (2009), Makos et al. (2009), Volkow et al. (2009), Andersen et al. (2010), Qu et al. (2010), Liu et al. (2012), Ueno et al. (2012), Sitaraman et al. (2015), Allada et al. (2017), Ly et al. (2018), Holst and Landolt (2018)
	Octopamine	Wake promoting	MB, parts of the central complex. PB. optic lobes regions of the ventral and lateral protocerebrum	Lee and Luo (2001), Sinakevitch and Strausfeld (2006), Crocker and Sehgal (2008), Busch et al. (2009), Crocker et al. (2010)
	Histamine	Wake promoting		Estelle and Simons (1994), Chu et al. (2004), Oh et al. (2013)
Function of neurotransmitters in promoting sleep	Serotonin	Sleep promoting	MB, FB	Witz et al. (1990), Saudou et al. (1992), Choi et al. (1995), Ursin (2002), Gao et al. (2002), Yuan et al. (2006), Haynes et al. (2015), Qian et al. (2017), Zhang et al. (2018), Gallopin et al. (2000), Mezler et al. (2001), Gong et al. (2004), Enell et al. (2007), Parisky et al. (2008), Okada et al. (2009), Peschel and Helfrich-Foerster (2011), Benedetto et al. (2012), Xiong et al. (2012), Li et al. (2013), Liu et al. (2014), Li et al. (2017)
	GABA	Sleep promoting	The sleep-promoting region of the hypothalamus and the median preoptic nucleus	

Dual functions of neurotransmitters	Glutamate	Wake promoting and sleep promoting	Widely express in the Drosophila central brain	Sinakevitch-Pean et al. (2001), DiAntonio (2006), Tomita et al. (2015), Zimmerman et al. (2017)
	Acetylcholine	Wake promoting and sleep promoting	PB, MB	Szerb, (1967), Buchner et al. (1986), Yasuyama and Salvaterra (1999), Salvaterra and Kitamoto (2001), Vazquez and Baghdoyan (2001), McCarthy et al. (2011), Yi et al. (2013), Van Dort et al. (2015), Xu et al. (2015), Holst and Landolt (2018)
Ion channels	Shaker	Sleep promoting	FB	Cirelli et al. (2005), Pimentel et al. (2016), Kim et al. (2020)
	Sandman	Wake promoting	FB	Yuan and Chen (2006)
Neuroendocrinic	Ecdysone	Sleep promoting	Adult mushroom bodies	Ishimoto and Kitamoto (2010)
	Juvenile hormones(JH)	Sexually dimorphic sleep	ND	Wu et al. (2018)
	Allatostatin-A (AstA)	Sleep promoting	R2 ring neurons of the ellipsoid body	Liu et al. (2016)
	Allatostatin-A (AstA)	Sleep promoting	dFB neurons	Donlea et al. (2018)
	Allatostatin-A (AstA)	Sleep promoting	PLP neurons and enteroendocrine cells	Chen et al. (2016)
	Myo inhibitory Peptide(allatostatin-B)	Sleep promoting	Arousal centre	Oh et al. (2014)

Continued

Table 1 Summary of sleep regulators in *Drosophila*.—cont'd

Type	Pathway/molecule	Effect on sleep	Functional tissues	References
	Drosophila insulin-like peptide (DILP)	Sleep promoting	1-LNvsand s-LNvs	Cong et al. (2015)
	Neuropeptide F (NPF)	Sleep promoting (Night)	Two groups of dorsal neurons (D1 and D2), two groups of dorsomedial neurons(P1 and P2), two groups of dorsolateral neurons (L1–1 and L1-s), one lateroventral group (L2), one subesophageal group (L2 and S) and the fan-shaped body (FB)	He et al. (2013)
	Neuropeptide F (NPF)	Wake promoting	Circadian pacemaker neurons	Chung et al. (2017)
	Short neuropeptide F (sNPF)	Wake promoting	Adult mushroom bodies	Chen et al. (2013)
Signalling pathways the regulates sleep	PKA/CREB pathway	Weak Promoting	Mushroom bodies	Cirelli et al. (1996), Mayford and Kandel (1999), Zamboni et al. (1999), Hendricks et al. (2001), Joiner et al. (2006), Chen et al. (2013), Kim et al. (2016), Maurer et al. (2016)
	G-protein Signalling	Sleep Promoting	Mushroom Bodies	Harmar et al. (2002), An et al. (2011), Guo et al. (2011), Yi et al. (2013), Doi et al. (2011), Doi et al. (2016), Huang et al. (2017)

	Ubiquitin-Proteasome pathway/hiw	Sleep Promoting	ND	Hunter-Ensor et al. (1996), Myers et al. (1996), Stanewsky et al. (1998), Naidoo et al. (1999), Lin et al. (2001), Koh et al. (2006), Stavropoulos and Young (2011)
	Ubiquitin-Proteasome pathway/Cul3	Sleep Promoting	ND	Seugnet et al. (2017)
	Ubiquitin-Proteasome pathway/faf	Wake promoting	ND	
	Unfolded Protein Response pathwavs/ JNK	Sleep promoting	Mushroom bodies	Ganguly-Fitzgerald et al. (2006), Naidoo (2009), Takahama et al. (2012), Singh et al. (2014), Brown et al. (2017), Bennett et al. (2018), Ly et al. (2020)
	Unfolded Protein Response pathwavs/ PERK	Sleep promoting	Mushroom bodies	
	Cell cycle signalling/ CycA	Sleep Promoting	PL	de Velasco et al. (2007), Rogulja and Young (2012), Afonso et al. (2015)
	Cell cycle signalling/ TARA	Sleep Promoting	PL	
	EGF pathway	Sleep Promoting	ND	Foltenyi et al. (2007), Fry et al. (2016), DeBardeleben et al. (2017)
	Notch signalling	Sleep homeostasis	Glial cell	Dobens et al. (2005), Alberi et al. (2011), Seugnet et al. (2011), Singh et al. (2011), Hayashi et al. (2016), Bennett et al. (2018)

Continued

Table 1 Summary of sleep regulators in *Drosophila*.—cont'd

Type	Pathway/molecule	Effect on sleep	Functional tissues	References
MicroRNAs	miR-190	Sleep promoting	ND	Goodwin et al. (2018)
	miR-955	Sleep promoting	ND	Goodwin et al. (2018)
	miR-956	Sleep promoting	ND	Goodwin et al. (2018)
	let-7	Sleep promoting	Mushroom body	Caygill and Johnston. (2008), Sokol et al. (2008), Chawla et al. (2016), Goodwin et al. (2018)
	miR-276a	Wake promoting	Clock neurons, PI, Mushroom body	Zhang et al. (2020)
	miR-276b	Wake promoting	Clock neurons, NPFR1 neurons, DopR1 neurons	Zhang et al. (2021)
	miR-375	Sleep Promoting	Clock neurons	Xia et al. (2020)
Protein translation regulators	BiP	Sleep promoting	Endoplasmic reticulum (ER)	Naidoo et al. (2007)
	eIF2a	Wake promoting	Endoplasmic reticulum (ER)	Axten et al. (2012), Ly et al. (2020)
	peIF2a	Sleep promoting	Endoplasmic reticulum (ER)	Ly et al. (2020)
	PERK	Sleep promoting	Endoplasmic reticulum (ER)	Ron and Walter (2007), Ly et al. (2020)
	Pumilio (pum)	Sleep promoting	The dorsal FB neurons	Weston and Baines (2007), De Jesus-Olmo et al. (2020)
Epigenetic factors	Tip60	Wake promoting (Day) sleep promoting (Night)	Small ventrolateral neurons (sLNvs)	Pirooznia et al. (2012)

	Stuxnet (Stx)	Wake promoting (Day)	Ellipsoid body	Zhao et al. (2021a,b)
Factors related to the sleep changes upon temperature shift	Shaker	Sleep promoting	GABA positive neurons	Kim et al. (2020)
	Extra sex combs (ESC)	Sleep promoting	Ellipsoid body and Mushroom body	Zhao et al. (2021a)
Factors important in mediating energy effect on sleep	Insulin like peptide (DILP)	Required for starvation-induced changes in sleep depth	ND	Brown et al. (2020)
	Adipokinetic hormone (AKH)	Promoting starvation-induced sleep loss	ND	He et al. (2020)
	Clock (Clk) and Cycle (Cyc)	Inhibiting starvation-induced sleep loss	Dorsally located Clk-expressing neurons	Keene et al. (2010)
	Translin	Promoting starvation-induced sleep loss	Leucokinin neurons	Murakami et al. (2016)
	NPF	Promoting starvation-induced sleep loss	ND	Chung et al. (2017)
	Leucokinin	Promoting starvation-induced sleep loss	A single pair of Lk neurons within the Lateral Horn (LHLK neurons)	Yurgel et al. (2019)

Continued

Table 1 Summary of sleep regulators in *Drosophila*.—cont'd

Type	Pathway/molecule	Effect on sleep	Functional tissues	References
Peripheral factors	Angiotensin-converting enzyme peptidase (ACER)	Wake promoting (Night)	Fat body	Carhan et al. (2011)
	Adipose triglyceride lipase brummer	Sleep promoting	Fat body	Grönke et al. (2005), Thimgan et al. (2010)
	Lipid storage droplet 2 (1sd2)	Sleep promoting	Fat body	Thimgan et al. (2010)
	Cytokine up aired 2 (upd2)	Wake promoting	Fat body	Rajan and Perrimon (2012)
	D-serine (D-Ser)	Sleep promoting	Gut	Dai et al. (2019)
Disease related factors	dHtt	Sleep promoting	ND	Gonzales and Yin (2010), Gonzales et al. (2016)
	LRRK2	Sleep promoting	ND	Sun et al. (2016)
	Human α-synuclein	Sleep promoting	ND	Ito et al. (2017)
	Parkin and pink1	Sleep fragmentation	ND	Valadas et al. (2018)
	Aβ42	Wake promoting	ND	Song et al. (2017)
	tau	Wake promoting	ND	Buhl et al. (2019)

Note: Some of the molecules may appear multiple times according to their functions. ND: not determined.

diseases also are summarized. These studies on the genetic control of sleep in *Drosophila* provide a great help in understanding the complex mechanisms of sleep regulation.

2. Circadian regulation of sleep

Circadian rhythm is controlled by about 150 "clock" neurons in the brain. Genes expressed in these neurons, including *clock(clk), cycle(cyc), period(per)* and *timeless(tim)*, form a central feedback loop regulating circadian oscillation of their own expression. Influences of clock neurons and associated genes on sleep regulation are presented here, while effects of other factors in the clock circuit on sleep regulation need to be further clarified.

Mutations of circadian genes have a profound effect on sleep. For example, pigment-dispersing factor *(Pdf)* deficiency mutations result in sleep increase (Parisky et al., 2008), while different alleles of *per* mutants affect sleep (Fropf et al., 2018). Moreover, clock genes are responsible for time-dependent sleep changes caused in various conditions. For example, the cyc^{01} mutation (loss of *cycle* function allele) abolished time-dependent sleep changes caused by the *210y*-Gal4 (the brain driver) driving neuronal activation to the sleep-promoting dorsal Fan-Shaped Body (Cavanaugh et al., 2014). Circadian gene mutations also result in homeostasis defects. Point mutations associated with loss of function alleles Per^{01} and Clk^{JRK} show a more prominent sleep rebound after sleep deprivation (a key index of homeostasis regulation of sleep) compared to Canton-S flies after 12h of sleep deprivation (Shaw et al., 2002). Interestingly, the *timeless* loss of function allele Tim^{0} exhibits disrupted sleep recovery after 3–6h of sleep deprivation (Hendricks et al., 2000; Shaw et al., 2002), but displays a sleep rebound similar to Per^{01} and Clk^{JRK} after 12h of sleep deprivation (Shaw et al., 2002). Moreover, cyc^{01} flies show the most exaggerated response to sleep loss, with much more sleep recovery compared to Per^{01} and Clk^{JRK} (Shaw et al., 2002). Neurons responsible for circadian rhythm are involved in sleep regulation. The lateral ventral neurons (LNvs) were shown to be wake promoting (Parisky et al., 2008; Shang et al., 2008; Sheeba et al., 2008), while the dorsal neurons 1 (DN1) are sleep promoting (Cavanaugh et al., 2014; Guo et al., 2016; Kunst et al., 2014). Communication between photoreceptors and clock neurons also is involved in sleep regulation (Damulewicz et al., 2020).

A diversity of genes expressed in clock neurons have been discovered to regulate sleep. For example, *resistant to dieldrin (Rdl)*, a $GABA_A$ receptor gene expressed in pdf neurons, has a role on promoting sleep (Parisky et al., 2008). The shaker cognate l *(Shal/Kv4)*, a well-known voltage-gated K channel in *Drosophila* circadian neurons, acts as a regulator of wake-sleep transition at dusk (Feng et al., 2018). Loss of *wide awake* (*wake*) in a set of arousal-promoting clock neurons, the large lateral ventral neurons (l-LNvs), exhibits a remarkable delay in sleep onset at dusk (Liu et al., 2014). The DN1 neuron expressing the diuretic hormone gene *DH31* secretes the neuropeptide gene product DH31 in response to the PDF signal to specifically regulate sleep, which acts as a wake promoting factor before dawn (Kunst et al., 2014).

Neural circuits of the circadian locomotor rhythm and sleep are connected. The *AstA*-secreting circadian pacemaker neurons directly synapse onto fan-shaped body (FB) neurons to promote sleep via glutamatergic synaptic transmission (Ni et al., 2019). A subset of dopaminergic neurons also was found to promote wakefulness during daytime in response to PDF (Potdar and Sheeba, 2018).

3. Synaptic signalling in wake/sleep regulation

Neurotransmitters play an important role in regulating sleep and wakefulness. So far, these include dopamine, octopamine, histamine, serotonin, GABA, acetylcholine and glutamate. Among them, dopamine, octopamine, and histamine promote the wake state, while serotonin and GABA promote sleep. Acetylcholine and glutamate are implicated as both wake and sleep-promoting factors (Ly et al., 2018).

3.1 Promotion of wake state

3.1.1 Dopamine

As an important neurotransmitter in the brain, dopamine (DA) is increasingly implicated in promotion of wake state. Around 127 dopaminergic neurons widely distributed in the *Drosophila* brain (Friggi-Grelin et al., 2003; Mao and Davis, 2009) project to the mushroom body (MB), fan-shaped body (FB), ellipsoid body (EB), noduli (NO), and protocerebral bridge (PB), with important roles in sleep regulation (Pitman et al., 2006). In *Drosophila*, enhancement of dopaminergic neurotransmission genetically or pharmacologically can dramatically reduce sleep and arousal threshold (Andretic et al., 2005; Holst and Landolt, 2018; Kume et al., 2005). Dopaminergic signalling in the MB mediates wake-promoting effects of caffeine (Andretic et al., 2008). In addition, a subset of MB-projecting DA neurons is specifically active during wake, and this activity can stimulate forgetting (Allada et al., 2017), providing a circuit basis for sleep-dependent memory.

Genetic deficiency mutations of dopamine signalling are known to promote arousal in *Drosophila*. Kume et al. reported a phenotypic characterization of the dopamine transporter (DAT) mutant known as *fumin (fmn)* by inserting a 43 bp fragment for loss of function, resulting in decrease of sleep (Kume et al., 2005). The DAT is responsible for dopamine presynaptic re-uptake. It was suggested that DAT loss-of-function in the *fmn* mutant prolongs dopamine signalling (Kume et al., 2005). Indeed, the *fmn* mutation actually impairs clearance of dopamine, thereby prolonging its action (Makos et al., 2009). Pharmacological studies found that 3-iodo-tyrosine (3IY), by inhibiting dopamine synthesis, causes sleep increases during daytime (Andretic et al., 2005). Consistent with this finding, the neurotoxin 6-hydroxydopamine ablates dopamine neurons in the grey matter around the ventral aqueduct of the midbrain, significantly reducing the amount of arousal (Lu et al., 2006). These findings provide further support for a role of dopamine in promoting arousal. It is worth noting that Positron Emission Tomography (PET) imaging in rhesus macaques and human subjects provides evidence that the wake-promoting drug modafinil increases levels of extracellular dopamine in the brain (Andersen et al., 2010; Volkow et al., 2009).

From fruit flies to mammals, the functional role of dopamine in regulating wakefulness is conserved. Consistent with the *Drosophila fmn* mutant phenotype, mice with impaired DAT exhibit elevated wakefulness (Qu et al., 2010), suggesting that sustained action of dopamine is important for induction of the wake state.

Dopamine receptors are important for regulation of arousal (Wisor et al., 2001). In vertebrates, two major classes of dopamine receptors (D_1-like receptors and D_2-like receptors) are recognized. D_1-like receptors include D_1 and D_5 receptors, whose activation increases levels of intracellular cyclic adenosine monophosphate (cAMP). On the other hand, activation of D_2-like receptors, including D_2, D_3 and D_4 subtypes, decreases cAMP levels (Missale et al., 1998). There are some notable structural differences between D_1-like receptors and D_2-like receptors (Missale et al., 1998). For example, D_1-like receptors rich in serine and threonine have much longer C-termini extending into the cytoplasmic compartment than D_2-like receptors. Besides, D2-like receptors have a longer third intracellular loop compared to D1-like receptors for coupling to different G proteins (Missale et al., 1998). It is important to note that D_2–like receptors have two main isoforms, the long isoform, D_{2L}, and the short isoform, D_{2S} (Missale et al., 1998). The D_{2L} differs from D_{2S} in having a 29-amino acid stretch in the third cytoplasmic loop, which does not affect ligand recognition but affects G-protein coupling and signal transmission. Pharmacological studies have found that the selective dopamine D_1 receptor agonist SKF38393, by intravenous or intraperitoneal administration to rabbits or rats, induces desynchronized electroencephalogram (EEG) waveforms and behavioural awakening (Isaac and Berridge, 2003).

Because of its relative simplicity compared to mammals, the *Drosophila* nervous system provides distinct advantages for relating neural circuits to behaviours. Nowadays, part of the wake-promoting dopamine signal has been isolated to single neurons in protocerebral posterior lateral 1 (PPL1) and protocerebral posterior medial region 3 (PPM3) clusters of the *Drosophila* brain that innervate neurons of the dorsal fan-shaped body (dFSB) in each hemisphere with dopamine D_1 receptors (Liu et al., 2012; Ueno et al., 2012). In addition, dopamine exerts wake-promoting effects outside of the dFSB. For example, dopaminergic activation of dopamine D_1 receptors in mushroom body neurons also induces wakefulness (Sitaraman et al., 2015).

3.1.2 Octopamine

Octopamine (OA) is a well-known congener of the vertebrate hormone norepinephrine. It was first discovered by Erspamer and Boretti (1951) in salivary gland extracts of Octopoda. Later, it was found that OA occurs in different tissues of invertebrates such as arthropods and molluscs. In insects, OA content is highest in the central nervous system (CNS) compared to other tissues. In the CNS of *Drosophila,* approximately 100 octopaminergic neurons send projections to many distinct regions in the brain, including the α', β' and γ regions of the mushroom bodies, parts of the central complex, the protocerebrum, as well as the optic lobes (Busch et al., 2009; Sinakevitch and Strausfeld, 2006).

Genetic and pharmacological manipulations of octopamine signalling demonstrate its arousal-promoting effects. Octopamine deficiency produced by loss of function mutations of *tyrosine decarboxylase 2* ($Tdc2^{RO54}$) or *tyramine β hydroxylase* ($T\beta H^{nm18}$) produced a phenotype of sleep increase, which was restored to the wild-type phenotype by pharmacological treatment with octopamine (Crocker and Sehgal, 2008). Consistent with these findings, electrical silencing of octopamine-producing cells decreases wakefulness, whereas excitation of these neurons promotes wakefulness (Crocker and Sehgal, 2008).

Octopamine acts via cAMP-dependent protein kinase (PKA) to reduce sleep. PKA activation plays an important role in sleep regulation of *Drosophila*. Indeed, overexpression of the regulatory subunit of PKA (*PKAr*) in adult neuronal tissue driven by the *Elav*GeneSwitch driver renders flies insensitive to sleep-reducing effects of octopamine, indicating that octopamine acts through a PKA-dependent pathway to promote arousal (Crocker and Sehgal, 2008).

At present, the neural circuitry for wake-promoting effects of octopamine has been characterized in considerable detail. Researchers utilized a mosaic analysis with a repressible cell marker (MARCM) to identify wake-promoting octopaminergic cells (Lee and Luo, 2001). They mapped a cluster of octopaminergic neurons in the medial protocerebrum, and found that these cells activate the mushroom body isoform of the octopamine receptor (OAMB) expressed by insulin-like-peptide (DILP2) neurons in the pars intercerebralis, depolarizing these cells and increasing their cAMP levels (Crocker et al., 2010).

3.1.3 Histamine

Histamine was considered to be an awakening substance as far back as 1970, but it did not attract attention until 1994, when a first-generation histamine H_1 receptor antagonist caused obvious sedative and hypnotic side effects in humans. This provided solid evidence that the histaminergic nervous system is involved in regulation of sleep-wake (Estelle and Simons, 1994).

Mammalian histaminergic neurons are distributed in the tuberomammillary nucleus (TMN) of the posterior hypothalamus, and they project throughout the brain. Release of histamine in the mammalian brain is positively correlated with wakefulness. Chu et al. (2004) used in vivo microdialysis combined with simultaneous electroencephalogram (EEG) and electromyogram (EMG) recordings to correlate sleep-wake with the amount of histamine released into the extracellular fluid of the frontal cortex of freely active rats (Chu et al., 2004). These studies showed that levels of histamine in the extracellular fluid of the rat during the awakening period are significantly higher relative to the sleep period. The amount released during awakening was four times that measured during sleep and was proportional to awakening time (Chu et al., 2004).

In addition to the important arousal function of histaminergic neurons in mammals, there is also evidence that histamine promotes arousal of *Drosophila* (Oh et al., 2013), where two histamine receptors have been described: the histamine gated chloride channel subunit 1 (HisCl1) and ora transientless (Ort). Hypomorphic deficiency mutations in histidine decarboxylase (HDC), an enzyme catalyzing conversion of histidine to histamine, caused increased sleep episode duration. Oh and colleagues found that the hypomorphic deletion mutations hdc^{P211} and hdc^{P218} showed a long-sleep phenotype. Pharmacological studies indicate that feeding histamine to *hisCl1* mutant flies can rescue wake to its wild-type state (Oh et al., 2013).

3.2 Promotion of sleep state

3.2.1 Serotonin

Serotonin (5-hydroxytryptamine, 5-HT) is a widely distributed sleep-promoting neurotransmitter in the *Drosophila* central nervous system (Yuan et al., 2006). Elevating serotonin levels either pharmacologically or genetically effectively enhances sleep in wild-type flies. Moreover, serotonin promotes sleep in some short-sleep mutants, suggesting that it can compensate for some sleep deficits (Yuan et al., 2006). Five serotonin GPCRs have been identified in the *Drosophila* genome, 5-HT1a, 5-HT1b, 5-HT2a, 5-HT2b and 5-HT7 (Choi et al., 1995; Qian et al., 2017; Saudou et al., 1992; Witz et al., 1990). In *Drosophila*, serotonin may regulate sleep through multiple receptors and brain regions to promote baseline sleep and homeostasis sleep rebound after sleep deprivation (Haynes et al., 2015; Qian et al., 2017; Yuan et al., 2006). Yuan and colleagues tested baseline sleep phenotypes of flies with genetically modified expression of three *Drosophila* serotonin receptors, and identified the 5-HT1a receptor as a promoter of sleep. Interestingly, the short and fragmented sleep phenotype of the 5-HT1a receptor deletion mutant can be rescued by its transgenic overexpression specifically in the adult mushroom body (Yuan et al., 2006). In addition, Haynes et al. (2015) and Qian et al. (2017) found that, in addition to influences of the 5-HT1a receptor, down-regulation of *tryptophan hydroxylase* (TRH) in the dorsal fan-shaped body projecting to the MBs and genetic knockout of the 5-HT2b receptor were sufficient to reduce sleep significantly. These results suggest that serotonergic signalling in specific neurons is required for regulation of sleep homeostasis.

In addition to *Drosophila*, serotonin also promotes sleep in zebrafish and mice. The raphe, a diffuse network of brainstem nuclei, synthesizes 5-HT and sends projections to almost every region of the brain (Oikonomou et al., 2019), is a conserved evolutionary structure of various organisms from fish to humans. By making and transporting serotonin to diverse brain regions, the raphe are critical for initiation and maintenance of sleep in zebrafish and mice (Oikonomou et al., 2019). EEG recordings showed that tryptophan hydroxylase 2(Tph2)-deficient mice, which cannot synthesize 5-HT in the brain, show decreased sleep (Zhang et al., 2018), indicating that 5-HT is important for sleep regulation. Due to the diverse serotonin receptor subtypes and brain regions involved, serotonergic signalling plays a variety of roles in promoting or inhibiting different stages of the sleep-wake cycle (Ursin, 2002). Interestingly, pharmacological studies have shown that serotonergic signalling affects sleep and wake in *Drosophila* (Gao et al., 2002). *Drosophila* models therefore may facilitate elucidation of mechanisms underlying serotonin-mediated sleep regulation in mammals.

3.2.2 GABA

The inhibitory transmitter γ-aminobutyric acid (GABA) is widely distributed in the central and peripheral nervous systems of *Drosophila*. Studies have shown that GABA plays important roles in initiation and maintenance of *Drosophila* sleep. The resistant to dieldrin gene *(Rdl)* expresses one of three $GABA_A$ subunits in the optic lobes, antennal lobes, mushroom bodies and central complex. Parisky et al. (2008) found that the loss of function $GABA_A$ receptor mutant ($RdlA^{302}$) exhibits decreased sleep. In addition, many molecular regulators of GABA signalling affect large ventral lateral neurons (l-LNvs), which are main components of the *Drosophila* circadian neural network (Peschel and Helfrich-Foerster, 2011). For example, activation of the protein WIDE AWAKE (WAKE) can increase *Rdl* expression at the end of day (Liu et al., 2014); the *Drosophila* Neuroligin 4 (DNlg4), a critical cell adhesion molecule, is required for proper clustering of Rdl in the l-LNvs (Li et al., 2013). In the clock cell, knocking out WAKE and DNIg4 in PDF-expressing neurons reduces the amount of sleep, indicating that WAKE and DNlg4 are both necessary for sleep regulation. However, the E3 ubiquitin ligase F-Box and Leucine Rich Repeat Protein 4 (Fbxl4) can degrade *Rdl* in clock cells, whereby knocking out *fbxl4* in PDF cells results in increases of sleep amount (Li et al., 2013). When *fbxl4* is degraded, GABA affects the wake state. In addition, GABA release from dorsal paired medial (DPM) neurons promotes sleep.

3.3 Dual functions of neurotransmitters

3.3.1 Glutamate

Glutamate is the major excitatory neurotransmitter at the *Drosophila* neuromuscular junction, where it plays a role similar to acetylcholine in mammals (DiAntonio, 2006). In addition, glutamate is widely expressed in the central brain of *Drosophila*, where it is involved in regulation of sleep and wakefulness (Sinakevitch-Pean et al., 2001). Previous studies indicate that glutamate signalling promotes wakefulness, especially at night (Zimmerman et al., 2017). By using the Gal4/UAS system to overexpress NaChBac (bacterial Na+ channel) or EKO (genetically modified Shaker K+ channel) in the brain, the activity of glutamatergic neurons in the adult brain was effectively activated or deactivated, respectively. Zimmerman et al. (2017) found that increased activity of glutamatergic neurons in the adult brain leads to significant increases in wakefulness at both daytime and nighttime, while decreased activity of these neurons reduces wakefulness at nighttime.

Previous data showed that glutamate signalling has a sleep-promoting effect on fruit flies (Tomita et al., 2015). Tomita et al. utilized the *fmn* deletion mutant to screen out the *N*-methyl-D-aspartic acid type glutamate receptor channel (NMDAR) gene, *Nmdar1*, which can promote sleep. If the *Nmdar1* was mutated or deleted, *Drosophila* sleep was significantly reduced. Pharmacological studies showed that flies fed the NMDAR antagonist MK801 exhibited a significant reduction in sleep. Together, these results suggest that NMDA glutamate receptors likely mediate sleep-promotion signalling in *Drosophila* under normal conditions (Tomita et al., 2015).

More recent research demonstrated that sleep homeostasis driven by sleep deprivation is regulated by glutamatergic neurons in the ellipsoid body of the fly brain (Liu et al., 2016). Extended wakefulness leads to increased expression of NMDA receptors in a subset of ellipsoid body neurons required for sleep rebound following prolonged wakefulness (Liu et al., 2016). Although glutamate expression promotes wakefulness as a whole, it seems that the location of glutamate release or postsynaptic glutamate receptors may determine whether glutamate signalling produces either sleep or wake-promoting signals.

3.3.2 Acetylcholine

The large ventrolateral neurons (l-LNvs) are primary circadian pacemakers in the *Drosophila* brain. Evidence shows that wakefulness is promoted by acetylcholine-mediated excitatory input to l-LNvs (McCarthy et al., 2011). Whole-cell recordings of l-LNv activity in vitro demonstrates that they fire with a synchronized rhythm and activation of their nicotinic acetylcholine receptors (nAChRs) increases excitation and firing (McCarthy et al., 2011). Pharmacological studies have shown that nicotinic receptor agonists depolarize l-LNvs, leading to action potential firing, while nAChR antagonists curare and α-bungarotoxin eliminate action potential firing (McCarthy et al., 2011). In summary, acetylcholine likely promotes wakefulness by exciting this part of the circadian circuitry in the *Drosophila* brain.

RNAi knockdown of the *vesicular acetylcholine transporter (vAChT)* to create acetylcholine deficiency in MB α/β core neurons significantly decreases sleep, suggesting that acetylcholine-mediated signalling from these cells promotes sleep (Yi et al., 2013). These sleep-promoting cholinergic neurons are neighbours to a layer of cholinergic wake-promoting cells, the surface/posterior and γ neurons. Interestingly, both sets of these neurons engage in inhibitory signalling via G_o. Consequently, it appears that neurons within the same brain region releasing the same neurotransmitter can have behaviourally opposing effects, possibly via a mutually inhibitory "flip-flop" sleep-switch model (Saper et al., 2010).

A second cluster of acetylcholine releasing neurons located in the basal forebrain (BF) project widely to the cortex (Holst and Landolt, 2018). Optogenetic activation of cholinergic PPT/LDT neurons in mice during non-rapid-eye-movement (NREM) sleep leads to a rapid transition into rapid-eye-movement (REM) sleep, but not into wakefulness (Van Dort et al., 2015). However, another recent study employing optogenetics corroborates evidence that acetylcholine neurons of the BF contribute to arousal by promoting wakefulness (Xu et al., 2015).

With the deepening of sleep research, more and more genes and neurotransmitters are implicated in regulation of sleep circuits, leading to a comprehensive understanding of the neuronal landscape in sleep regulation. Although the *Drosophila* brain has a relatively simple nervous system, regulation of sleep has many similarities with mammals. At present, areas of the fly brain including the mushroom body, dorsal fan-shaped body, pars intercerebralis and lateral ventral neurons have been demonstrated to have important roles for sleep regulation, in which the dorsal fan-shaped body primarily promotes sleep, while the mushroom body contains both sleep and wake-promoting neurons.

3.4 Ion channels

Ion channels provide the basis for electrical signalling in brain neurons and consequently have an indispensable role in maintaining various life activities, including sleep-wake. The *shaker (Sh)* gene encodes a voltage-gated potassium channel (K^+) controlling membrane repolarization and transmitter release. Loss-of-function *shaker* mutant flies have an overall increased neuronal excitability and a short-sleeping phenotype. Cirelli et al. screened 9000 mutant fly lines and identified *minisleep (mns)*, a line exhibiting one-third the level of sleep observed in wild-type flies (Cirelli et al., 2005). They found that *mns* flies carry a point mutation in a conserved domain of the *shaker* gene (Cirelli et al., 2005). Kim et al. demonstrated that *shaker*-expressing GABAergic neurons projecting onto the dorsal fan-shaped body regulate temperature-adaptive sleep behaviours in *Drosophila*. Under low temperature conditions, *shaker* mutant flies sleep less, while light and high temperature cooperatively impact *shaker* effects on sleep (Kim et al., 2020). Further research showed that the *Rdl* gene expressed in dFSB neurons exerts a negative influence downstream of *shaker* (Kim et al., 2020). Voltage-gated potassium channels (Kv) are also involved in mammalian sleep control. Mice lacking the closest mammalian homologue of *shaker* (*Kv1.2*) sleep less than wild-type mice. However, the mouse phenotype is less dramatic than that observed in *Drosophila*, probably because of the redundancy of mammalian *Kv* genes (Yuan and Chen, 2006).

In the dorsal fan-shaped body of *Drosophila*, the potassium channel *sandman* acts as a wake-promoting ion channel. *Sandman* is encoded by the CG8713 gene and translocates to the plasma membrane in response to dopamine, while it in turn suppresses activity of dFSB neurons to increase arousal (Pimentel et al., 2016). Therefore, *sandman* appears to function as a switch to turn off sleep by regulating excitability of sleep-controlling dFSB neurons.

4. Neuroendocrine regulation of sleep

4.1 Ecdysone

Steroid hormones regulate a variety of developmental, physiological and behavioural processes. Recent research has demonstrated that 20-hydroxyecdysone (20E) mediated signalling in the adult is intimately involved in transitions between physiological states of sleep and wakefulness (Ishimoto and Kitamoto, 2010). Four lines of evidence suggest that ecdysone is involved in homeostatic sleep regulation. First, 20E promotes sleep of adult flies in a dose-dependent manner by altering the length of sleep and wake bouts, but without affecting waking activity. Second, heterozygous mutants for ecdysone synthesis display a short-sleep phenotype, which can be rescued by treatment with 20E. Third, heterozygous mutants for nuclear ecdysone receptors showed reduced sleep, while overexpression of wild-type ecdysone receptors in the adult mushroom body resulted in an isoform-specific sleep increase. Finally, endogenous ecdysone levels increased after sleep deprivation, and mutants with defective ecdysone signalling showed little sleep rebound. It is reasonable to hypothesize that shared mechanisms underlie both the transitions between different developmental stages and transitions between behavioural states in adults, both of which are regulated by ecdysone signalling (Ishimoto and Kitamoto, 2010).

4.2 Allatostatin

Sleep-promoting allatostatin neurons send projections to the dorsal fan-shaped body (dFB) of the central complex and act as a feedback controller or homeostat (Donlea et al., 2014). dFB neurons induce sleep via a range of inhibitory signals that include the neuropeptide allatostatin-A (AstA) (Donlea et al., 2018). Targets of AstA contain a group of central complex interneurons named helicon cells. These neurons are inhibited by the sleep-promoting AstA in response to visual input, transmitting to R2 ring neurons of the ellipsoid body, whose activity is associated with the accumulation of sleep need (Liu et al., 2016).

In addition, specific thermogenetic activation of AstA-positive PLP (posterior lateral protocerebrum) neurons and enteroendocrine cells promotes sleep and reduces feeding in the fruit fly. PLP neurons, which express a functional PDF receptor, function as a down-stream target of PDF. Silencing of AstA signalling by tethering PDF changes the sleep/activity ratio in opposite directions, but does not affect rhythmicity (Chen et al., 2016).

In addition, AstC is a novel signalling neuropeptide expressed in DN1, DN3, and LPN clock neuron clusters in the *Drosophila*, which inhibits a single LNd clock cell to modulate evening locomotor activity under short (winter-like) and long (summer-like) day photoperiods. Expression of AstC-R2 in LNds promotes a similar phenotype. Since the AstC/AstC-R2 signalling pathway co-participates in rhythm regulation, it may also play an important role in sleep regulation (Diaz et al., 2019).

Myoinhibitory Peptide (MIP), also known as allatostatin-B or prothoracicostatic peptide, is expressed in central interneurons with little sexual dimorphism. The sex peptide receptor (SPR), for which MIP is a ligand, mediates MIP actions to modulate neural activities of the arousal system involving PDF neurons. Downregulation of MIP in the brain results in decreased sleep and mutants lacking MIP fail to maintain normal sleep homeostasis. Sleep deprivation facilitates MIP secretion from a small subset of brain neurons and axonal processes innervating dendritic fields of the PDF neurons. So, the MIP-SPR signalling pathway may function as a sleep homeostat that senses the need for sleep and stabilizes sleep by providing a slow-acting inhibitory input to an arousal centre (Oh et al., 2014).

4.3 Juvenile hormones

Sesquiterpenoid juvenile hormones (JHs) play important roles in reproduction, moulting and insect growth. During larval stages, JHs maintain juvenile phenotypes by preventing metamorphic actions of the moulting hormone ecdysone (Yokoi et al., 2020).

Recent research reveals that JH signalling regulates sexual dimorphism of sleep, in which male and female adults exhibit long- and short-sleep phenotypes, respectively. Blocking JH signalling pathways can eliminate these differences. Germ cell-expressed (GCE), which encodes one of the fly JH receptors, clearly mediates sexually dimorphic sleep phenotypes, since the *gce* deletion mutant cannot be rescued by JH treatment. Sexually dimorphic sleep induced by JH signalling is closely related to sexual genes. Besides, JH also plays an important role in antagonism of aging-induced sleep reduction in male flies (Wu et al., 2018).

4.4 Insulin-like peptide

Drosophila insulin-like peptides (DILPs) are important neuroendocrine signalling molecules. Seven DILPs have been identified in *Drosophila*, whose functions in metabolism and aging are highly conserved. DILPs and their receptor DInR are demonstrated to work also in regulation of sleep behaviour, whereby deletion mutants of DILPs or the DInR exhibit significant reduction of total sleep, whereas over-expression of DILPs or the DInR in the nervous system leads to increased sleep (Cong et al., 2015).

In addition, both starvation and a sucrose-only diet result in sleep increases, suggesting that dietary protein is essential for normal sleep and homeostasis. Furthermore, Dilp2 is involved in sleep responses to starvation, in which Dilp2 levels decrease significantly when flies are starved (Cong et al., 2015). Flies lacking Dilp2 exhibit a compensatory sleep rebound following starvation-induced sleep deprivations. Together, these findings reveal innate resilience to starvation-induced sleep loss and underlie starvation-induced changes in sleep duration and sleep depth (Brown et al., 2020).

4.5 Neuropeptide F

Drosophila neuropeptide F (NPF) is a homologue of mammalian neuropeptide Y (NPY). NPF and its receptor NPFR regulate feeding behaviour of insects in response to food quality (Brown et al., 1999; Mercer et al., 2011). Recently, NPF signalling has been implicated in regulation of sleep. Overexpression of either NPF or NPFR1 produced increased total sleep during nighttime with increasing episode duration, and less sleep rebound after sleep deprivation (He et al., 2013) via promotion of downstream ecdysone signalling (Kannangara et al., 2020). As mentioned above, ecdysone also plays roles in sleep regulation, probably through involvement in a complex signalling network. Besides, the NPF network forms a critical component in regulation of both adult sleep and feeding. Activation of NPF signalling in the brain promotes wakefulness and adult feeding, likely through its cognate receptor NPFR (Chung et al., 2017).

NPF neurons expressing the circadian photoreceptor cryptochrome (CRY) are largely responsible for changes in sleep behaviour, while feeding is considered to be controlled by the NPF cells in the midgut in *Drosophila*, demonstrating the diverse functional roles played NPF cells in regulation of wakefulness and hunger.

4.6 Short neuropeptide F

Another neuroendocrine signal in *Drosophila* is short neuropeptide F (sNPF), which is involved in regulation of food intake, response to stress, neuronal excitability and circadian rhythm. However, sNPF also regulates sleep (Chen et al., 2013; He et al., 2013; Shang et al., 2013). Loss of function mutants of sNPF or its receptor sNPFR1 produce significant increases in sleep bout duration. Sleep deprivation results in increased sNPF levels in wild-type control flies, accompanied by wakefulness at nighttime. However, deletion of the sNPF gene makes flies resistant to wakefulness caused by sleep deprivation, suggesting that sNPF is important in sleep homeostasis (Chen et al., 2013). Furthermore, sNPF signalling negatively regulates sleep through the cAMP/PKA/CREB signalling pathway (Chen et al., 2013). Flies experience negative sleep rebound upon cessation of sNPF neuron activity (Shang et al., 2013). In addition, activation of sNPF neurons produces a homeostatic sleep drive that can be dissipated long after neuronal activity ceases (Juneau et al., 2019).

5. Signalling pathways that regulate sleep

5.1 PKA/CREB pathway

The cAMP/PKA/CREB (cAMP response element binding protein) signalling pathway is associated with learning and memory in all studied organisms (Mayford and Kandel, 1999). The relationship between sleep and memory consolidation has attracted intense interest. Hendricks et al. suggested that the cAMP/PKA/CREB pathway is particularly important in regulating the rest-activity cycle (Hendricks et al., 2001). They found that the activities of the cAMP and the CREB are negatively related to sleep, which was further demonstrated by Chen et al. (2013). Up-regulation of the cAMP/PKA/CREB pathway leads to decreased sleep, while its down regulation is associated with sleep increase (Chen et al., 2013). Activity of CREB is negatively correlated with sleep amount (Hendricks et al., 2001), indicating that the cAMP/PKA/CREB pathway is negatively correlated with sleep demand.

Dramatic effects on sleep are due to increased activity of cAMP-dependent protein kinase (PKA) in space and time in adult mushroom bodies (Joiner et al., 2006). PKA overexpression under control of an inducible panneural promoter is sufficient to reduce sleep. Use of specific promoters to express PKA in different regions of the brain led to the demonstration that sleep phenotypes were most significant when expression was targeted to the mushroom body (MB) (Joiner et al., 2006). Use of inducible promoters aimed at electrical silencing via potassium channel expression increased sleep, while increasing excitability through NaChBac expression reduced sleep (Joiner et al., 2006). Therefore, PKA can increase sleep by inhibiting output of neurons in the MB.

Since CREB is a transcription factor involved in sleep regulation, some genes affected by the CREB might be expected to be the wake-promoting genes. Light is one of the strongest environmental factors for entraining endogenous circadian rhythms. In *Drosophila*, the role of CREB-regulated transcription co-activator (CRTC) in maintaining circadian rhythm is independent of light. Deletion of CRTC significantly reduces light-independent oscillation of the TIMELESS (TIM) protein in clock neurons, and leads to a long, but poor locomotory rhythm in continuous darkness (Kim et al., 2016). Previous studies confirmed that the CREB binding protein (CBP) plays an important role in activating transcription of clock genes for regulation of the circadian rhythm, in which moderate increase of CBP levels strongly inhibits circadian rhythms without affecting the molecular oscillations of circadian transcription (Maurer et al., 2016).

In mammals, cAMP signalling and CREB activity are associated with waking up (Cirelli et al., 1996). However, cAMP accumulation decreases during sleep deprivation and is replenished by sleep, which is related to the area of the hypothalamus during recovery (Zamboni et al., 1999).

5.2 G-protein signalling

G-protein-coupled receptors (GPCRs) are the most abundant receptor family in both vertebrates and invertebrates. There are about 40 types of G proteins, most of which are associated with the cell membrane and which consist of three different subunits named α, β and γ, with a total molecular weight of about 100 kDa. The classification is based on differences in structure, properties of the α subunit, and evolutionary similarity. G proteins are divided into six categories: Gs, Gi, Go, Gq, Gt, and G?, whereby G? is a designation for G proteins of undetermined functions. Guo et al. demonstrated that Go signalling is very important in sleep maintenance. Indeed, increasing Go signalling leads to sleep consolidation in *Drosophila* (Guo et al., 2011). Targeted activation of Go in the MB promotes sleep, while inhibition of endogenous Go signalling in this part of the brain leads to sleep loss. Further experiments demonstrate that sleep regulation is mediated by Go independently of the cAMP pathway (Guo et al., 2011). Yi et al. (2013) found that inhibition of Go through over-expression of pertussis toxin (PTX) in the MB driven by c309-Gal4 leads to a dramatic sleep increase at daytime and fragmented sleep at nighttime. They narrowed the brain region involved to the cholinergic MB α/β core neurons, which are responsible for Go signalling-mediated sleep effects (Yi et al., 2013).

G-protein signalling also plays an important role in mammalian sleep (Doi et al., 2011, 2016). The regulator of G-protein signalling 16 (RGS16), a known inactivated G protein in mammals, selectively increases circadian rhythm dependence of intracellular cAMP signals in the suprachiasmatic nucleus (SCN). RGS16-deficient flies exhibit a lengthening of the circadian period. The clock-controlled RGS16 can precisely adjust cAMP signalling to maintain the normal phase relationship between dorsal and ventromedial SCN activity. Therefore, RGS16-dependent time regulation of the intracellular G protein signal coordinates intercellular synchronization of SCN pacemaker neurons, thus defining the 24-h behavioural rhythm (Doi et al., 2011). According to the type of G protein, many downstream signalling pathways can be activated. Circadian fluctuation of cAMP signalling is key to maintaining circadian clock function in the SCN. In this context, the vasoactive intestinal polypeptide (VIP) receptor 2 (Vipr2) is a positive regulator of cAMP (An et al., 2011), and has been proved to be necessary for SCN time-keeping (Harmar et al., 2002). G-protein-coupled receptor 176 (Gpr176) is a unique orphan GPCR mainly in the SCN, and its protein abundance fluctuates according to a circadian rhythm. Gpr176 can co-localize with and act independently of Vipr2. Gpr176 as a repressor has an agonist-independent basal activity to repress cAMP production. On the contrary, Vipr2 is an activator of cAMP. It should be noted that the activity of Gpr176 requires a unique G protein subclass Gz rather than the standard Gi (Doi et al., 2016). In *C. elegans,* G-protein signalling also plays an important role in sleep regulation (Huang et al., 2017).

5.3 Ubiquitin-proteasome pathway

Entrainment of the *Drosophila* clock is achieved by light-induced degradation of TIM (Hunter-Ensor et al., 1996; Myers et al., 1996). Light is transmitted to TIM by the circadian photoreceptor cryptochrome (CRY) (Stanewsky et al., 1998). Both CRY and TIM are degraded by the proteasome (Lin et al., 2001; Naidoo et al., 1999), in which the circadian photoreceptor CRY responds to light and transmits signals to TIM, which is degraded by the ubiquitin-proteasome pathway and E3 ligase JET (Koh et al., 2006). CRY itself is also degraded by light, but with a slower time course than that of TIM (Koh et al., 2006). Through forward genetic screening of *Drosophila*, Stavropoulos and Young identified a gene named insomniac (*inc*), mutations of which cause significant reduction in total sleep due to a decrease at sleep episode duration (Stavropoulos and Young, 2011). They also showed *inc* expression is independent of the circadian clock (Stavropoulos and Young, 2011). Furthermore, *inc* has a physiological relationship with Cullin-3 (Cul3), a ubiquitin ligase complex, in which decreased Cul3 activity in neurons reproduces the phenotype of insomnia. The relationship between *inc* and Cul3 suggests that the ubiquitin-proteasome pathway may play a universal role in controlling sleep and arousal in animals (Stavropoulos and Young, 2011). The E3 ubiquitin ligase gene *HighWire* (*hiw*) and *fat facet* (*faf*) involved in ubiquitin-proteasome signal transduction have recently been identified as sleep regulators. *Hiw* can promote sleep and increase Cul3 activity, while *faf* acting as a deubiquitination enzyme is a wake-promoting molecule. Furthermore, increasing expression of *faf* or/and decreasing *hiw* in l-LNvs neurons induces sleep loss (Seugnet et al., 2017). These studies provide ample evidence that the ubiquitin-proteasome pathway modulates sleep in *Drosophila*.

5.4 Cellular stress/unfolded protein response pathways

There are many stress systems in cells, among which endoplasmic reticulum (ER) stress is an important form. When cells are stimulated by internal and external factors, the equilibrium state of ER morphology and function is destroyed, protein processing and transportation are blocked, and a large number of unfolded or misfolded proteins accumulate in the ER. In this case, cells will accordingly respond and relieve ER stress to promote recovery of function.

Accordingly, stress-induced accumulation of unfolded or misfolded proteins can result in the "unfolded protein response" (UPR). Immunoglobulin heavy chain-binding protein (BiP), an ER molecular chaperone, is one of the essential components in this process. Studies on BiP indicate that the UPR mechanism is involved in modulation of sleep recovery after sleep deprivation. Levels of BiP mRNA and protein are upregulated in the *Drosophila* brain after sleep deprivation, which is accompanied by an increase in sleep recovery (Naidoo et al., 2007). The BIP protein is a key indicator of ER stress, which helps to determine the duration of sleep recovery following sleep deprivation (Naidoo et al., 2007). In *Drosophila*, social isolation (SI) leads to reduced sleep (Ganguly-Fitzgerald et al., 2006). Reports by Brown et al. connected for the first time social isolation and the unfolded protein response (UPR) (Brown et al., 2017). Ly et al. have shown that PERK in the UPR signal transduction pathway is necessary for normal sleep. Knockdown of the PERK signal in neurons inhibits sleep, while its over-expression significantly increases sleep. In addition, PERK activity in a small group of wake-up neurons is sufficient to alter sleep and wake-up behaviour, as well as expression of the wake-up neuropeptide pigment-dispersing factor (PDF) (Ly et al., 2020).

At the cellular level, UPR triggers three adaptive and protective cellular responses: (1) Up-regulation of ER chaperones, such as BIP/GRP78; (2) attenuation of protein translation; (3) ER-associated degradation (ERAD) of misfolded proteins by the proteasome. These three reactions reduce protein load and ER stress. When UPR-mediated cytoprotection fails, CHOP (C/EBP homologous protein), c-Jun amino terminal kinase

(JNK) and caspases are involved in the apoptotic response to sustained ER stress (Naidoo, 2009). JNK is involved in the insulin pathway and regulates metabolism and longevity. In *Drosophila*, JNK silencing in insulin producing neurons of the brain has a slight effect on sleep. Panneuronal knockdown of JNK by RNA interference leads to declining sleep and life span. In addition, reducing JNK expression in the mushroom body has a significant effect on sleep (Takahama et al., 2012). JNK-1 is a conserved activator of stress responses across species. In *C. elegans,* loss of degradation function by ERAD can lead to increased total sleep (Bennett et al., 2018; Singh et al., 2014).

5.5 Cell cycle signalling

Proteins that regulate cell cycle signalling also play a role in *Drosophila* sleep. Through large-scale screening of UAS-RNAi fly lines, Rogulja and Young showed that knockdown of the Rca1, a regulator of cyclin A1 (CycA) and a homologue of a mammalian early mitotic inhibitor 1, results in decreased sleep time (Rogulja and Young, 2012). Reducing the abundance of CycA in neurons prolongs the transition from wakefulness to sleep, causes multiple sleep arousals, and reduces the homeostatic response to sleep deprivation (Rogulja and Young, 2012). Another forward genetic study found that Taranis (TARA), a *Drosophila* homologue of the Trip-Br (SERTAD) family of transcriptional co-regulators, regulates sleep-like behaviour (Afonso et al., 2015). TARA acts on the *Drosophila* brain pars lateralis (PL), considered to be a functional analogue of the mammalian hypothalamus (de Velasco et al., 2007). Knocking down TARA in the PL reduces sleep significantly (Afonso et al., 2015). In addition, TARA regulates CycA protein levels by interacting with the *CycA* gene and promotes sleep through cyclin-dependent kinase 1 (*Cdk1*) activity (Afonso et al., 2015). Therefore, it is reasonable to conclude that cell cycle dynamics may be involved in sleep regulation.

5.6 Inducing factor pathways

5.6.1 EGF pathway

Epidermal growth factor receptor (EGFR) signalling is a sleep-promoting pathway in *Drosophila* (Foltenyi et al., 2007). Sleep increases in a dose-dependent mode during nighttime or daytime after over-expression of the processing proteins Rho and Star, while sleep is reduced after Rho-RNAi expression. Furthermore, Rho and Star increase sleep through activation of extracellular signal-regulated kinase (ERK) (Foltenyi et al., 2007).

The role of EGF in sleep appears to be conserved in other species. In *C. elegans*, epidermal growth factor receptor (EGFR) signalling is required in the ALA interneuron to promote sleep-like behavioural quiescence after cellular stress (Fry et al., 2016). Stress-induced sleep (SIS) is triggered when epidermal growth factor (EGF) activates the neurosecretory ALA neuron, which then releases neuropeptides to promote sleep. SIS in *C. elegans* is important for restoration of cellular homeostasis (DeBardeleben et al., 2017).

5.7 Notch signalling

While the Notch signalling pathway has been intensively studied in the context of cell differentiation, it is increasingly evident that it plays important roles in adult nervous system function as well. In addition, considering that Notch signalling plays a role in synaptic plasticity (Alberi et al., 2011; Hayashi et al., 2016), a popular theoretical hypothesis is that it may be one of the molecular signals involved in the effects of sleep and stress on synaptic plasticity.

Notch signalling appears to play a role in regulation of sleep homeostasis (Seugnet et al., 2011). These authors found that levels of the transcription factor *bunched* (*bun*), a negative regulator of Notch signalling (Dobens et al., 2005), was upregulated by sleep deprivation. Seugnet et al. (2011) found that over-expression of the Notch ligand DELTA in the mushroom body, a structure involved in learning, memory and sleep recovery, reduces sleep rebound. Similarly, expression of a Notch functional allele also causes decreased sleep rebound. To sum up, these data suggest that sleep deprivation produces sleep rebound via suppression of the Notch signal through upregulation of *bunched*. Studies have shown that the Notch pathway also regulates sleep in *C. elegans*, where disturbance of the Notch signal can change sleep amount and depth of sleep; these changes nevertheless are complex. With the increase of reaction time, the Notch signal will increase the amount of sleep and lead to deep sleep (Singh et al., 2011). Reduced function of the Notch ligand LAG-2 or JNK-1 kinase results in increased total sleep during development (Bennett et al., 2018).

6. Other molecules

6.1 MicroRNAs

MicroRNAs (miRNAs) are non-coding regulators of mRNA translation. Consisting of 22–25 nucleotides typically, they usually exert a negative influence on translation through binding to mRNAs in conjunction with the RISC complex (Huntzinger and Izaurralde, 2011). Some miRNAs also regulate the circadian clock (Chen et al., 2014; Vodala et al., 2012).

To find microRNAs involved in sleep regulation, Goodwin et al. used a library of "miRNA sponge-expressing" flies to carry out a genetic screen, and discovered that about 75% of sleep-regulating miRNAs are required in neurons, and different miRNAs play specific roles in sleep, in which some lead to sleep decrease, while others cause sleep increase. Furthermore, miRNAs regulate total sleep either through regulation of sleep episode number or sleep episode duration (Goodwin et al., 2018).

The miRNA *let-7* is proposed to increase both daytime and nighttime sleep. The *Drosophila let-7* gene is located in a cluster with two other miRNAs, *mir-100* and *mir-125* (Caygill and Johnston, 2008; Chawla et al., 2016), to comprise the "*let-7* complex" (Sokol et al., 2008). To confirm the influence of *let-7*, CRISPR/Cas9 was used to disrupt the 5′-seed sequence of *let-7*, while maintaining *mir-100* and *mir-125* sequences intact (Goodwin et al., 2018). Down regulation of *let-7* (UAS-*let-7* RNAi) driven by VT030559-Gal4 (a strong pan-MB driver) led to decreases of total sleep and the conclusion that *let-7* promotes sleep (Goodwin et al., 2018).

To more clearly understand the global functional miRNA-mRNA interaction network involved in the circadian system, Xia et al. (2020) used CLEAR (Covalent Ligation of Endogenous Argonaute-bound RNAs)-CLIP (Cross-Linking and Immuno-Precipitation) to explore regulatory functions of miRNAs in the circadian system. They compared miRNA-mRNA interactions between the *Drosophila* wild-type strain W^{1118} and a mutant of the key circadian transcriptional regulator Clock (Clk^{jrk}), thereby identifying tens of thousands of miRNA-mRNA interactions in both the head and body. In this way, they identified miR-375, a novel regulator of sleep, and found that it targets *timeless* to promote sleep (Xia et al., 2020). Besides miR-375, additional miRNAs and related targets, such as miR-1-timeless, miR-1-vri, miR-9c-clock, also were identified (Xia et al., 2020).

Drosophila has two microRNA-276 paralogs, miR-276a and miR-276b, differing by only one nucleotide. Zhang et al. found that these miRNAs, regulated by the core CLOCK/CYCLE (CLK/CYC) transcription factor upstream, negatively regulate sleep (Zhang et al., 2020, 2021). Although miR-276a and miR-276b have similar phenotypes, their seed regions activated by the core CLOCK are different, and their expression patterns in brain and down-stream targets differ. For example, miR-276a is widely expressed in the mushroom body (MB), pars intercerebralis (PI) and lateral dorsal clock neurons (LNds), while miR-276b has a broader distribution in MB, PI, LNd, lateral ventral neurons (LNv), fan-shaped body (FB) and dorsal neurons (DN). Furthermore, miR-276a regulates sleep via suppression of TIM and NPFR1 (The neuropeptide F receptor), while miR-276b is essential for maintaining sleep by targeting TIM, NPFR1 and DopR1 (Dopamine 1-like receptor 1) (Zhang et al., 2020, 2021). In addition to the above, other miRNAs are implicated in sleep regulation, in which miR-190 is required for promotion of nighttime sleep, while miR-956 and miR-955 promote daytime sleep (Goodwin et al., 2018).

6.2 Protein translation regulation

6.2.1 Perk

Clinical and animal experiments indicate that sleep disorders may be related to destruction of protein homeostasis in the brain, for example, through accumulation of misfolded protein aggregates (Kurtishi et al., 2019). Protein kinase R (PKR)-like endoplasmic reticulum kinase (PERK) regulates protein balance in cells through transient inhibition of protein synthesis. To maintain protein homeostasis, cells coordinate protein synthesis by folding and appropriate intracellular transport (Ron and Walter, 2007). If protein homeostasis is destroyed, misfolded proteins accumulate and an intracellular signalling pathway called unfolded protein response (UPR) is activated (Ron and Walter, 2007). UPR originates in the endoplasmic reticulum (ER), and its activation is mediated by three signalling transducers: Protein kinase R (PKR)-like endoplasmic reticulum kinase (PERK), inositol-requiring protein-1 (IRE1), and activating transcription factor6 (ATF6). UPR further activates signalling cascades to inhibit protein synthesis, upregulate protein chaperone activity, and degrade transcript and protein products together in the cell (Ron and Walter, 2007).

UPR signalling is involved in sleep homeostasis via the UPR chaperone molecule binding immunoglobulin protein (BiP) (Naidoo et al., 2007), which assists in assembly of intracellular molecules and protein folding. In *Drosophila*, genetic overexpression of BiP increases recovery sleep after sleep deprivation, while overexpression of the dominant-negative BiP leads to a reduction in recovery sleep (Naidoo et al., 2007). Ly et al. examined changes in phosphorylated eIF2a (*peIF2a*) in the PERK pathway, and found that *pelF2a* increases following sleep deprivation, accomplished through slow and continuous mechanical movement to promote wakefulness (Ly et al., 2020). Ly et al. also discovered that administration of GSK2606414, a small molecule inhibitor of PERK, results in a higher proportion of actively translating polysomes in the brain, and treatment of animals with the small molecule drug ISRIB represses the PERK pathway via activation of eIF2B, which reduces nighttime sleep time (Axten et al., 2012). These data suggest that the UPR is required not only for sleep homeostasis but also for normal sleep. Since pharmacological administration of PERK has no spatial specificity in all cell types, the role of the PERK on sleep behaviour was evaluated through a genetic approach. PERK knockout results in sleep decrease, while its overexpression induces more sleep by regulating translation of proteins via elF2a phosphorylation (Ly et al., 2020). PERK activity in a small subset of wake-promoting neurons is sufficient to alter *Drosophila* sleep and wake behaviour and change expression of the wake-promoting neuropeptide PDF (Ly et al., 2020).

6.2.2 *Pumilio*

Pimentel et al. (2016) pointed out that dorsal FB neurons seem to manipulate homeostatic sleep, and that they are electrically active during wake, while electrically silent during sleep. This study suggests that neuronal excitability is an important effector of sleep regulation. The *Pumilio* (*pum*) gene, whose main function is to inhibit protein translation, is indispensable for compensatory sleep after sleep deprivation in *Drosophila* (De Jesus-Olmo et al., 2020). In this instance, neuronal excitability is regulated by repressing translation of the voltage-gated sodium channel (Weston and Baines, 2007). Because of its central role in regulating membrane excitability, sodium channel repression by *pum* is essential for regulation of neuronal homeostasis. De Jesus-Olmo et al. pointed out that regulation of homeostatic sleep rebound is independent of baseline sleep (De Jesus-Olmo et al., 2020). In addition, *pum* also regulates expression of the eukaryotic initiation factor 4E (eIF4E), which likely affects sleep through its influence on synaptic bouton size and number, synaptic growth and function (Cao et al., 2010; Menon et al., 2004; Vessey et al., 2006).

6.3 Epigenetic factors

6.3.1 *Histone modification factors are involved in sleep control in* Drosophila

Tip60, a histone acetyltransferase (HAT), regulates sleep in *Drosophila* through mediation of axonal growth of small lateral ventral neurons (sLNvs), which serve as pacemaker cells (Pirooznia et al., 2012). The function of Tip60 is mediated via production of the neuropeptide pigment-dispersing factor (PDF) to stabilize *Drosophila* sleep-wake cycles.

Stuxnet (*Stx*), a Polycomb stabilization factor, influences sleep through the *Stx-Polycomb- Octβ2R* regulatory cascade in the ellipsoid body. Mutation of *Stx* results in up-regulation of daytime sleep. *Stx* positively regulates the *Octβ2R* through suppression of *Polycomb*. More interestingly, *Stx* is octopamine-responsive, dependent upon the *Octβ2R*. In response to octopamine, *Octβ2R* transcription is inhibited as a result of *stuxnet* down-regulation. The Stuxnet-Polycomb-*Octβ2R* cascade provides a feedback mechanism for OA signals to the ellipsoid body (EB) to regulate sleep homeostasis and sleep amount. It provides a brake, eliminating the unfavourable effect caused by excessive OA signalling (Zhao et al., 2021b).

6.3.2 Genetic factors related to the sleep changes upon temperature shift

Balance between circuits containing GABAergic neurons and dopaminergic neurons is responsible for temperature effects on sleep. This was discovered through analysis of *shaker* (*Sh*) and *Hyperkinetic* (*Hk*) deletion mutants that display short-sleep phenotypes in a temperature-sensitive manner (Kim et al., 2020). Further investigation indicates that *Sh* depletion in GABAergic neurons partially phenocopies Sh mutants. Under low temperature conditions, dopaminergic input to the dFSB is inhibited, which results in sleep promotion. Under high temperature conditions, GABAergic synaptic input to the dFSB is inhibited, which results in a wake-promoting effect (Kim et al., 2020).

Extra sex combs (ESC) mediated *Hsp* regulation is responsible for buffering multiple sleep-related stresses (Zhao et al., 2021a,b). ESC is a component of the PcG (Polycomb group) complex, which proved to be an important epigenetic regulator of many biological processes such as development, diseases, and so on. Under the condition of temperature shift, daytime and nighttime sleep structure is reconstructed. Changing temperature from 25 °C to 29 °C results in increased daytime and decreased nighttime sleep of control flies (Parisky et al., 2016). Interestingly, ESC deletion flies exhibit a much more drastic restructuring of sleep, which could be rescued by *Hsps* mutations (Zhao et al., 2021a,b). Moreover, the *esc* mutation leads to a higher survival rate upon sleep deprivation, while *Hsp68*, *Hsp70A* and *Hsp70Ba* mutant flies are more susceptible to sleep deprivation stress (Zhao et al., 2021a,b).

6.4 Genetic factors are important in mediating energy effect on sleep

As is well known, starvation leads to less sleep (Keene et al., 2010). The *Drosophila* insulin-like peptide (DILP) potentially responds to energy signals. DILP positive cells in the pars intercerebralis (PI) respond to octopamine to suppress sleep (Crocker et al., 2010). Recent study showed that adipokinetic hormone (AKH) is important for transmission of starvation signals to sleep-controlling neurons. AKH is similar to mammalian glucagon regarding maintenance of energy balance through enhancement of catabolism and mobilization of stored energy (Gade et al., 1997; Sajwan et al., 2015). As for locomotory behaviour, previous reports showed that elimination of AKH-producing cells decreases locomotor activity (Isabel et al., 2005), while facilitation of AKH release by treatment with tolbutamide increases daytime activity of flies and decreases daytime sleep (Metaxakis et al., 2014). A recent study pointed out that starvation-induced sleep loss occurs in the first transformation of photoperiod either from light to dark or from dark to light, independent of starvation time. AKH receives and transmits the starvation signal to the Fork head Box-O (FOXO) gene, and then adjusts sleep through regulation of axonal remodelling of s-LNv dorsal projections (He et al., 2020).

In addition to DILPs and AKH signalling, *clock/cycle* (Keene et al., 2010), *Translin* (Murakami et al., 2016), NPF (Chung et al., 2017), serine (Sonn et al., 2018), and *Leucokinin* (Yurgel et al., 2019) were found to regulate starvation-induced sleep loss in *Drosophila. Clock* and *cycle* limit starvation-induced sleep loss (Keene et al., 2010), and blocking serine metabolism enhances starvation-induced sleep loss (Sonn et al., 2018), while loss of function of *translin*, *NPF*, or *leucokinin* reduces hunger-induced sleep loss (Chung et al., 2017; Murakami et al., 2016; Yurgel et al., 2019).

6.5 Regulation of sleep by factors outside of the nervous system

Adipose tissue modulates behaviours through controls on biosynthesis and metabolism of energy (Morton and Schwartz, 2011). As the centre of energy biosynthesis and metabolism, the fat body is very important in regulation of numerous behaviours, including courtship, feeding and egg-laying (Lazareva et al., 2007; Sassu et al., 2012; Xu et al., 2011). Angiotensin-converting enzyme peptidase (*acer*), a sleep-regulating gene, is highly expressed in the fat body. Mutations of *acer* disrupt nighttime sleep, raising the likelihood that fat body metabolism is important in sleep regulation

(Carhan et al., 2011). Several genes are expressed preferentially in the fat body, including adipose triglyceride lipase *brummer*, whose mutation elevates triglyceride stores and enhances homeostatic sleep response to sleep deprivation (Grönke et al., 2005; Thimgan et al., 2010). Conversely, fly mutants lacking *lipid storage droplet 2* (*lsd2*) exhibit reduced triglyceride levels and lack of homeostatic sleep rebound following sleep deprivation. These data suggest that triglyceride stores in the fat body are essential for the homeostatic response to sleep deprivation (Thimgan et al., 2010). Besides, adipose tissue may regulate sleep through secretion of hormones. In mammals, leptin binds to hypothalamic receptors to modulate sleep and feeding (Inutsuka and Yamanaka, 2013). The male ob/ob mouse mutant, a genetic model of leptin deficiency, was found to disrupt sleep architecture significantly with an elevated level of arousal (Laposky et al., 2006). An orthologue of mammalian leptin, *cytokine unpaired 2* (*upd2*), was identified in *Drosophila* (Rajan and Perrimon, 2012). Secretion of *upd2* from the fat body can regulate insulin release from Insulin Producing Cells (IPCs) (Rajan and Perrimon, 2012). These findings suggest that the *upd2* may function through IPCs to regulate sleep; this remains to be verified.

Sleep disorders are associated with intestinal physiology in humans and animals (Everson and Wehr, 1993; Hasler et al., 2004). The intestine senses environmental conditions and communicates with the central nervous system (Haber et al., 2017). In mammals and flies, crosstalk between endogenous endocrine cells and neurons through neuropeptide signalling regulates processes such as energy homeostasis (Droujinine and Perrimon, 2016; Han et al., 2011). Recently, the *N*-methyl-D-aspartate glutamate receptor (NMDAR) and D-serine (D-Ser) have been implicated in sleep regulation in both flies and mammals (Liu et al., 2016; Papouin et al., 2017; Tomita et al., 2015). D-Ser produced by intestinal serine racemase promotes sleep through NMDAR1 signalling in *Drosophila* (Dai et al., 2019). Sleep deprivation may lead to accumulation of ROS in the fly gut. Accumulated ROS are causally linked to decreased survival (Vaccaro et al., 2020). In summary, influences of the peripheral nervous system on sleep regulation are significant and underestimated, highlighting the need for further investigation.

6.6 Disease molecules affecting sleep

Researchers have constructed a number of *Drosophila* models to address relationships between gene expression and human diseases. *Drosophila* disease models mainly address mechanisms of Huntington disease (HD), Parkinson's disease (PD), Alzheimer's disease (AD) and a range of cancer syndromes. The first three diseases are neurodegenerative, caused by neuropathy, while cancers are caused by abnormal metabolism and proliferation of cells. The most widely used model for HD is the R6/2 mouse model, which has a circadian rhythm defect (Pallier and Morton, 2009). However, mice are nocturnal animals and exhibit different sleep and activity patterns from humans. *Drosophila melanogaster* is a low-light creature whose activity pattern is similar to humans with a fixed rest pattern at night (Cirelli, 2009; Gerstner and Yin, 2010; Shaw et al., 2000).

Gonzales and Yin showed that sleep in multiple *Drosophila* HD models is significantly reduced during nighttime, and sleep fragmentation is obvious. Knockdown of the endogenous *Drosophila* homologue of the huntingtin (*dHtt*) gene is sufficient to disrupt nocturnal sleep (Gonzales and Yin, 2010), illustrating its role in sleep maintenance. However, its internal mechanism remains unknown. Through expression of full length *dHtt* and N-terminally truncated *mutHtt* pan-neuronally, Gonzales et al. (2016) found that sleep initiation is impaired, sleep amount is reduced, and abnormal nocturnal activity is induced. Sleep deprivation in the HD model results in destruction of sleep homeostasis with aggravated sleep deficiency, along with changes of the PKA/CREB signal pathway (Gonzales et al., 2016).

Compared to HD, PD is more common. It is the second most common neurodegenerative disease, affecting nearly 1% of the world's population over the age of 65 (Braak et al., 2003; Dauer and Przedborski, 2003). Sleep disorder is not only a main feature of the disease, but also a secondary effect resulting from drug treatment, tremor, stiffness and motor dyskinesia. Major sleep disorders include excessive drowsiness during daytime and rapid eye movement sleep behaviour disorder (RBD). RBD may precede motor symptoms for many years (Boeve et al., 1998). Using the PD fly model, Seugnet et al. found that sleep deprivation in middle stages of the pathology results in a persistent short-term memory impairment for more than 3 days (Seugnet et al., 2009). Although the pathogenicity of PD is not clear, it is generally believed that genetic susceptibility and environmental factors are the main contributors. The leucine-rich repeat kinase 2 (LRRK2) gene mutation is the most common genetic cause of familial and the sporadic PD (Cookson, 2010; Greggio and Cookson, 2009; Paisán-Ruíz et al., 2004). Sleep disorder is the main symptom of patients with the LRRK2 mutation (Dawson et al., 2010). Using the PD *Drosophila* model, Sun et al. (2016) demonstrated that expression of hLRRK2 in the MB causes sleep problems and presynaptic dysfunction. Melatonin treatment attenuates hLRRK2-induced sleep disorder and synaptic dysfunction, highlighting its potential therapeutic potential (Sun et al., 2016). Feeding flies the oriental traditional herbal medicine Kamikihito and the Unkei-to improves PD sleep deficiency significantly. Unkei-to also can improve motor function in PD flies (Ito et al., 2017). Valadas et al. (2018) have found that abnormal contacts of ER and mitochondria in PD mutants reduces secretion of phosphatidylserine from ER and disrupts production of neuropeptide-containing vesicles. Feeding phosphatidylserine to PD mutant *Drosophila* rescues production of peptidergic vesicles and quickly restores the normal sleep pattern, indicating that excessive endoplasmic reticulum-mitochondrial contact contributes to sleep and rhythm disorders (Valadas et al., 2018).

AD is another common neurodegenerative disease in humans, for which researchers have constructed useful *Drosophila* models (UAS strains of Aβ and tau). AD is a complex disease associated with β-amyloid (Aβ) peptide processing abnormalities, including tau protein hyperphosphorylation, inflammation, oxidative damage, decrease of neurotrophic factors, changes in the balance between excitatory and inhibitory synapses, and cognitive impairment leading to dementia (John and Berg, 2015; Li et al., 2016). Common symptoms of AD involve impaired sleep patterns, in which the amount of sleep is reduced and fragmentation of sleep is pronounced. Dissel et al. found using the *Drosophila* AD model that sleep can reverse cognitive impairment (Dissel et al., 2017). Accumulation of Aβ42 in AD flies results in a significant decrease in sleep, accompanied by the axonal denitrification after development of PDF neurons. The altered dendritic structure is mediated by JNK activation and can be reversed by reducing the JNK signal (Song et al., 2017). Buhl focused on tau-induced sleep disorders in the AD model, in which overexpression of tau in *tim* neurons significantly reduces the amount of sleep and induces abnormal activity (Buhl et al., 2019).

In addition to the aforementioned neurodegenerative diseases, *Drosophila* has been widely used in research on human cancer in recent years, and a large number of cancer models have been constructed. According to tissue classification, *Drosophila* models can be divided into gut cancer model (Apidianakis and Rahme, 2011; Biteau and Jasper, 2011; Wang et al., 2014), brain cancer model (Yuan et al., 2015), angiogenesis model (Affolter et al., 2009; Grifoni et al., 2015), lung cancer model (Andrew and Ewald, 2010; Levine and Cagan, 2016), prostate cancer model (Ito et al., 2014; Wilson et al., 2017), thyroid cancer model (Levine and Cagan, 2016) and liquid tumour model (Krzemień et al., 2007; Osman et al., 2009; Sinenko et al., 2010). Clinical data show that cancer patients in different stages of disease often show various degrees of sleep disorder. However, mechanisms of sleep disorders associated with different cancers vary significantly, providing intriguing directions for further exploration in the future.

7. Conclusions and future directions

Genetic approaches to sleep research have resulted in considerable progress during the past 20 years. To summarize, studies in *Drosophila* have made huge contributions to our understanding of how gene expression regulates sleep. To date, molecules identified in sleep regulation fall into a variety of categories, including neurotransmitters, circadian genes, ion channels, endocrine molecules and molecules involved in signalling pathways. In addition, sleep is a quantifiable, phenotypic character that is controlled by many genes. Sleep studies require a variety of different approaches in order to obtain a holistic view of the regulatory network. Recently, approaches via quantitative genetics have been applied to sleep research, in which some key genes for sleep regulation have been discovered (Allebrandt et al., 2017; Harbison et al., 2013, 2017; Serrano Negron et al., 2018).

In the future, systematic mapping of neurons and neural circuits controlling sleep in *Drosophila* using screens involving Gal4, CRISPR/Cas9 and other genetic tools will be crucial for further advances in our understanding. To construct a comprehensive picture of molecular mechanisms of sleep, forward genetic screens or genome-wide sequencing techniques will be applied. Furthermore, investigation and application of a new generation of live-imaging tools will contribute to mechanistic analyses of how genes programme sleep regulation.

Acknowledgements

We thank Prof. Michael Adams (University of California, Riverside) for revision of this manuscript. The work was supported by the National Natural Science Foundation of China Grants 31970458 to ZZ and 32070492 to JD.

References

Affolter, M., Zeller, R., Caussinus, E., 2009. Tissue remodelling through branching morphogenesis. Nat. Rev. Mol. Cell Biol. 10, 831–842.

Afonso, D.J., Liu, D., Machado, D.R., Pan, H., Jepson, J.E., Rogulja, D., Koh, K., 2015. TARANIS functions with cyclin A and Cdk1 in a novel arousal center to control sleep in drosophila. Curr. Biol. 25, 1717–1726.

Alberi, L., Liu, S., Wang, Y., Badie, R., Smith-Hicks, C., Wu, J., Pierfelice, T.J., Abazyan, B., Mattson, M.P., Kuhl, D., Pletnikov, M., Worley, P.F., Gaiano, N.,

2011. Activity-induced notch signaling in neurons requires arc/Arg3.1 and is essential for synaptic plasticity in hippocampal networks. Neuron 69, 437–444.
Allada, R., Cirelli, C., Sehgal, A., 2017. Molecular mechanisms of sleep homeostasis in flies and mammals. Cold Spring Harb. Perspect. Biol. 9, a027730.
Allebrandt, K.V., Teder-Laving, M., Cusumano, P., Frishman, G., Levandovski, R., Ruepp, A., Hidalgo, M.P.L., Costa, R., Metspalu, A., Roenneberg, T., De Pittà, C., 2017. Identifying pathways modulating sleep duration: from genomics to transcriptomics. Sci. Rep. 7, 4555.
An, S., Irwin, R.P., Allen, C.N., Tsai, C., Herzog, E.D., 2011. Vasoactive intestinal polypeptide requires parallel changes in adenylate cyclase and phospholipase C to entrain circadian rhythms to a predictable phase. J. Neurophysiol. 105, 2289–2296.
Andersen, M.L., Kessler, E., Murnane, K.S., McClung, J.C., Tufik, S., Howell, L.L., 2010. Dopamine transporter-related effects of modafinil in rhesus monkeys. Psychopharmacology (Berl) 210, 439–448.
Andretic, R., van Swinderen, B., Greenspan, R.J., 2005. Dopaminergic modulation of arousal in drosophila. Curr. Biol. 15, 1165–1175.
Andretic, R., Kim, Y.-C., Jones, F.S., Han, K.-A., Greenspan, R.J., 2008. Drosophila D1 dopamine receptor mediates caffeine-induced arousal. Proc. Natl. Acad. Sci. U. S. A. 105, 20392–20397.
Andrew, D.J., Ewald, A.J., 2010. Morphogenesis of epithelial tubes: insights into tube formation, elongation, and elaboration. Dev. Biol. 341, 34–55.
Apidianakis, Y., Rahme, L.G., 2011. Drosophila melanogaster as a model for human intestinal infection and pathology. Dis. Model. Mech. 4, 21–30.
Axten, J.M., Medina, J.R., Feng, Y., Shu, A., Romeril, S.P., Grant, S.W., Li, W.H.H., Heerding, D.A., Minthorn, E., Mencken, T., Atkins, C., Liu, Q., Rabindran, S., Kumar, R., Hong, X., Goetz, A., Stanley, T., Taylor, J.D., Sigethy, S.D., Tomberlin, G.H., Hassell, A.M., Kahler, K.M., Shewchuk, L.M., Gampe, R.T., 2012. Discovery of 7-Methyl-5-(1-{ 3-(trifluoromethyl)phenyl acetyl}-2,3-dihydro-1H-indol-5 -yl)-7H-pyrrolo 2,3-d pyrimidin-4-amine (GSK2606414), a potent and selective first-in-class inhibitor of protein kinase R (PKR)-like endoplasmic reticulum kinase (PERK). J. Med. Chem. 55, 7193–7207.
Benedetto, L., Chase, M.H., Torterolo, P., 2012. GABAergic processes within the median preoptic nucleus promote NREM sleep. Behav. Brain Res. 232, 60–65.
Bennett, H.L., Khoruzhik, Y., Hayden, D., Huang, H., Sanders, J., Walsh, M.B., Biron, D., Hart, A.C., 2018. Normal sleep bouts are not essential for *C. elegans* survival and FoxO is important for compensatory changes in sleep. BMC Neurosci. 19, 10.
Biteau, B., Jasper, H., 2011. EGF signaling regulates the proliferation of intestinal stem cells in Drosophila. Development 138, 1045–1055.
Boeve, B.F., Silber, M.H., Ferman, T.J., Kokmen, E., Smith, G.E., Ivnik, R.J., Parisi, J.E., Olson, E.J., Petersen, R.C., 1998. REM sleep behavior disorder and degenerative dementia: an association likely reflecting Lewy body disease. Neurology 51, 363–370.
Braak, H., Del Tredici, K., Rüb, U., de Vos, R.A., Jansen Steur, E.N., Braak, E., 2003. Staging of brain pathology related to sporadic Parkinson's disease. Neurobiol. Aging 24, 197–211.
Brown, M.R., Crim, J.W., Arata, R.C., Cai, H.N., Chun, C., Shen, P., 1999. Identification of a Drosophila brain-gut peptide related to the neuropeptide Y family. Peptides 20, 1035–1042.
Brown, M.K., Strus, E., Naidoo, N., 2017. Reduced sleep during social isolation leads to cellular stress and induction of the unfolded protein response. Sleep 40, zsx095.
Brown, E.B., Shah, K.D., Faville, R., Kottler, B., Keene, A.C., 2020. Drosophila insulin-like peptide 2 mediates dietary regulation of sleep intensity. PLoS Genet. 16, e1008270.

Buchner, E., Buchner, S., Crawford, G., Mason, W.T., Salvaterra, P.M., Sattelle, D.B., 1986. Choline Acetyltransferase-like immunoreactivity in the brain of Drosophila Melanogaster. Cell and Tissue Res. 246, 57–62.

Buhl, E., Higham, J.P., Hodge, J.J.L., 2019. Alzheimer's disease-associated tau alters Drosophila circadian activity, sleep and clock neuron electrophysiology. Neurobiol. Dis. 130, 104507.

Busch, S., Selcho, M., Ito, K., Tanimoto, H., 2009. A map of octopaminergic neurons in the Drosophila brain. J. Comp. Neurol. 513, 643–667.

Cao, Q., Padmanabhan, K., Richter, J.D., 2010. Pumilio 2 controls translation by competing with eIF4E for 7-methyl guanosine cap recognition. RNA 16, 221–227.

Carhan, A., Tang, K., Shirras, C.A., Shirras, A.D., Isaac, R.E., 2011. Loss of angiotensin-converting enzyme-related (ACER) peptidase disrupts night-time sleep in adult Drosophila melanogaster. J. Exp. Biol. 214, 680–686.

Cavanaugh, D.J., Geratowski, J.D., Wooltorton, J.R., Spaethling, J.M., Hector, C.E., Zheng, X., Johnson, E.C., Eberwine, J.H., Sehgal, A., 2014. Identification of a circadian output circuit for rest: activity rhythms in Drosophila. Cell 157, 689–701.

Caygill, E.E., Johnston, L.A., 2008. Temporal regulation of metamorphic processes in Drosophila by the let-7 and miR-125 heterochronic microRNAs. Curr. Biol. 18, 943–950.

Chawla, G., Deosthale, P., Childress, S., Wu, Y.-c., Sokol, N.S., 2016. A let-7-to-miR-125 MicroRNA switch regulates neuronal integrity and lifespan in Drosophila. PLoS Genet. 12, e1006247.

Chen, W.F., Shi, W., Li, L.Z., Zheng, Z., Li, T.J., Bai, W.W., Zhao, Z.W., 2013. Regulation of sleep by the short neuropeptide F (sNPF) in Drosophila melanogaster. Insect Biochem. Mol. Biol. 43, 809–819.

Chen, W., Liu, Z., Li, T., Zhang, R., Xue, Y., Zhong, Y., Bai, W., Zhou, D., Zhao, Z., 2014. Regulation of Drosophila circadian rhythms by miRNA let-7 is mediated by a regulatory cycle. Nat. Commun. 5, 5549.

Chen, J.T., Reiher, W., Hermann-Luibl, C., Sellami, A., Cognigni, P., Kondo, S., Helfrich-Forster, C., Veenstra, J.A., Wegener, C., 2016. Allatostatin A signalling in Drosophila regulates feeding and sleep and is modulated by PDF. PLoS Genet. 12, e1006346.

Choi, D.S., Loric, S., Colas, J.F., Callebert, J., Rosay, P., Kellermann, O., Launay, J.M., Maroteaux, L., 1995. The mouse 5-HT2B receptor: homologous subtype or species variant? Behav. Brain Res. 73, 253–257.

Chu, M., Huang, Z.L., Qu, W.M., Eguchi, N., Yao, M.H., Urade, Y., 2004. Extracellular histamine level in the frontal cortex is positively correlated with the amount of wakefulness in rats. Neurosci. Res. 49, 417–420.

Chung, B.Y., Ro, J., Hutter, S.A., Miller, K.M., Guduguntla, L.S., Kondo, S., Pletcher, S.D., 2017. Drosophila neuropeptide F signaling independently regulates feeding and sleep-wake behavior. Cell Rep. 19, 2441–2450.

Cirelli, C., 2009. The genetic and molecular regulation of sleep: from fruit flies to humans. Nat. Rev. Neurosci. 10, 549–560.

Cirelli, C., Pompeiano, M., Tononi, G., 1996. Neuronal gene expression in the waking state: a role for the locus coeruleus. Science 274, 1211–1215.

Cirelli, C., Bushey, D., Hill, S., Huber, R., Kreber, R., Ganetzky, B., Tononi, G., 2005. Reduced sleep in Drosophila shaker mutants. Nature 434, 1087–1092.

Cong, X.N., Wang, H.L., Liu, Z.X., He, C.X., An, C.J., Zhao, Z.W., 2015. Regulation of sleep by insulin-like peptide system in Drosophila melanogaster. Sleep 38, 1075–1089.

Cookson, M.R., 2010. The role of leucine-rich repeat kinase 2 (LRRK2) in Parkinson's disease. Nat. Rev. Neurosci. 11, 791–797.

Crocker, A., Sehgal, A., 2008. Octopamine regulates sleep in Drosophila through protein kinase A-dependent mechanisms. J. Neurosci. 28, 9377–9385.

Crocker, A., Shahidullah, M., Levitan, I.B., Sehgal, A., 2010. Identification of a neural circuit that underlies the effects of octopamine on sleep: wake behavior. Neuron 65, 670–681.

Dai, X., Zhou, E., Yang, W., Zhang, X., Zhang, W., Rao, Y., 2019. D-serine made by serine racemase in Drosophila intestine plays a physiological role in sleep. Nat. Commun. 10, 1986.

Damulewicz, M., Ispizua, J.I., Ceriani, M.F., Pyza, E.M., 2020. Communication among photoreceptors and the central clock affects sleep profile. Front. Physiol. 11, 993.

Dauer, W., Przedborski, S., 2003. Parkinson's disease: mechanisms and models. Neuron 39, 889–909.

Dawson, T.M., Ko, H.S., Dawson, V.L., 2010. Genetic animal models of Parkinson's disease. Neuron 66, 646–661.

De Jesus-Olmo, L.A., Rodriguez, N., Francia, M., Aleman-rios, J., Pacheco-agosto, C.J., Ortega-torres, J., Nieves, R., Fuenzalida-uribe, N., Ghezzi, A., Agosto, J.L., 2020. Pumilio regulates sleep homeostasis in response to chronic sleep deprivation in Drosophila melanogaster. Front. Neurosci. 14, 319.

de Velasco, B., Erclik, T., Shy, D., Sclafani, J., Lipshitz, H., McInnes, R., Hartenstein, V., 2007. Specification and development of the pars intercerebralis and pars lateralis, neuroendocrine command centers in the Drosophila brain. Dev. Biol. 302, 309–323.

DeBardeleben, H.K., Lopes, L.E., Nessel, M.P., Raizen, D.M., 2017. Stress-induced sleep after exposure to ultraviolet light is promoted by p53 in Caenorhabditis elegans. Genetics 207, 571–582.

DiAntonio, A., 2006. Glutamate receptors at the Drosophila neuromuscular junction. Int. Rev. Neurobiol. 75, 165–179.

Diaz, M.M., Schlichting, M., Abruzzi, K.C., Long, X., Rosbash, M., 2019. Allatostatin-C/AstC-R2 is a novel pathway to modulate the circadian activity pattern in Drosophila. Curr. Biol. 29, 13–22.

Dissel, S., Klose, M., Donlea, J., Cao, L., English, D., Winsky-Sommerer, R., van Swinderen, B., Shaw, P.J., 2017. Enhanced sleep reverses memory deficits and underlying pathology in Drosophila models of Alzheimer's disease. Neurobiol. Sleep Circadian Rhythms 2, 15–26.

Dobens, L., Jaeger, A., Peterson, J.S., Raftery, L.A., 2005. Bunched sets a boundary for notch signaling to pattern anterior eggshell structures during Drosophila oogenesis. Dev. Biol. 287, 425–437.

Doi, M., Ishida, A., Miyake, A., Sato, M., Komatsu, R., Yamazaki, F., Kimura, I., Tsuchiya, S., Kori, H., Seo, K., Yamaguchi, Y., Matsuo, M., Fustin, J.M., Tanaka, R., Santo, Y., Yamada, H., Takahashi, Y., Araki, M., Nakao, K., Aizawa, S., Kobayashi, M., Obrietan, K., Tsujimoto, G., Okamura, H., 2011. Circadian regulation of intracellular G-protein signalling mediates intercellular synchrony and rhythmicity in the suprachiasmatic nucleus. Nat. Commun. 2, 327.

Doi, M., Murai, I., Kunisue, S., Setsu, G., Uchio, N., Tanaka, R., Kobayashi, S., Shimatani, H., Hayashi, H., Chao, H.W., Nakagawa, Y., Takahashi, Y., Hotta, Y., Yasunaga, J., Matsuoka, M., Hastings, M.H., Kiyonari, H., Okamura, H., 2016. Gpr176 is a Gz-linked orphan G-protein-coupled receptor that sets the pace of circadian behaviour. Nat. Commun. 7, 10583.

Donlea, J.M., Pimentel, D., Miesenbock, G., 2014. Neuronal machinery of sleep homeostasis in Drosophila. Neuron 81, 860–872.

Donlea, J.M., Pimentel, D., Talbot, C.B., Kempf, A., Omoto, J.J., Hartenstein, V., Miesenböck, G., 2018. Recurrent circuitry for balancing sleep need and sleep. Neuron 97, 378–389.e374.

Droujinine, I.A., Perrimon, N., 2016. Interorgan communication pathways in physiology: focus on Drosophila. Annu. Rev. Genet. 50, 539–570.

Enell, L., Hamasaka, Y., Kolodziejczyk, A., Nassel, D.R., 2007. g-aminobutyric acid (GABA) signaling components in Drosophila: immunocytochemical localization of GABA(B) receptors in relation to the GABA(A) receptor subunit RDL and a vesicular GABA transporter. J. Comp. Neurol. 505, 18–31.

Erspamer, V., Boretti, G., 1951. Substances of a phenolic and indolic nature present in acetone extracts of the posterior salivary glands of octopoda (Octopus vulgaris, Octopus macropus and Eledone moschata). Experientia 7, 271–273.

Estelle, F., Simons, R., 1994. H-1-receptor antagonists-comparative tolerability and safety. Drug Saf. 10, 350–380.

Everson, C.A., Wehr, T.A., 1993. Nutritional and metabolic adaptations to prolonged sleep deprivation in the rat. Am. J. Physiol. 264, R376–R387.

Feng, G., Zhang, J., Li, M., Shao, L., Yang, L., Song, Q., Ping, Y., 2018. Control of sleep onset by Shal/K(v)4 channels in Drosophila circadian neurons. J. Neurosci. 38, 9059–9071.

Foltenyi, K., Andretic, R., Newport, J.W., Greenspan, R.J., 2007. Neurohormonal and neuromodulatory control of sleep in Drosophila. Cold Spring Harb. Symp. Quant. Biol. 72, 565–571.

Friggi-Grelin, F., Coulom, H., Meller, M., Gomez, D., Hirsh, J., Birman, S., 2003. Targeted gene expression in Drosophila dopaminergic cells using regulatory sequences from tyrosine hydroxylase. J. Neurobiol. 54, 618–627.

Fropf, R., Zhou, H., Yin, J.C.P., 2018. The clock gene period differentially regulates sleep and memory in Drosophila. Neurobiol. Learn. Mem. 153, 2–12.

Fry, A.L., Laboy, J.T., Huang, H., Hart, A.C., Norman, K.R., 2016. A conserved GEF for rho-family GTPases acts in an EGF signaling pathway to promote sleep-like quiescence in Caenorhabditis elegans. Genetics 202, 1153–1166.

Gade, G., Hoffmann, K.-H., Spring, J.H., 1997. Hormonal regulation in insects: facts, gaps, and future directions. Physiol. Rev. 77, 963–1032.

Gallopin, T., Fort, P., Eggermann, E., Cauli, B., Luppi, P.H., Rossier, J., Audinat, E., Muhlethaler, M., Serafin, M., 2000. Identification of sleep-promoting neurons in vitro. Nature 404, 992–995.

Ganguly-Fitzgerald, I., Donlea, J., Shaw, P.J., 2006. Waking experience affects sleep need in Drosophila. Science 313, 1775–1781.

Gao, J., Zhang, J.X., Xu, T.L., 2002. Modulation of serotonergic projection from dorsal raphe nucleus to basolateral amygdala on sleep-waking cycle of rats. Brain Res. 945, 60–70.

Gerstner, J.R., Yin, J.C., 2010. Circadian rhythms and memory formation. Nat. Rev. Neurosci. 11, 577–588.

Gonzales, E., Yin, J., 2010. Drosophila models of Huntington's disease exhibit sleep abnormalities. PLoS Curr 2, RRN1185.

Gong, H., McGinty, D., Guzman-Marin, R., Chew, K.T., Stewart, D., Szymusiak, R., 2004. Activation of c-fos in GABAergic neurones in the preoptic area during sleep and in response to sleep deprivation. J. Physiol-London 556, 935–946.

Gonzales, E.D., Tanenhaus, A.K., Zhang, J., Chaffee, R.P., Yin, J.C., 2016. Early-onset sleep defects in Drosophila models of Huntington's disease reflect alterations of PKA/CREB signaling. Hum. Mol. Genet. 25, 837–852.

Goodwin, P.R., Meng, A., Moore, J., Hobin, M., Fulga, T.A., Van Vactor, D., Griffith, L.C., 2018. MicroRNAs regulate sleep and sleep homeostasis in Drosophila. Cell Rep. 23, 3776–3786.

Greggio, E., Cookson, M.R., 2009. Leucine-rich repeat kinase 2 mutations and Parkinson's disease: three questions. ASN Neuro 1, e00002.

Grifoni, D., Sollazzo, M., Fontana, E., Froldi, F., Pession, A., 2015. Multiple strategies of oxygen supply in Drosophila malignancies identify tracheogenesis as a novel cancer hallmark. Sci. Rep. 5, 9061.

Grönke, S., Mildner, A., Fellert, S., Tennagels, N., Petry, S., Müller, G., Jäckle, H., Kühnlein, R.P., 2005. Brummer lipase is an evolutionary conserved fat storage regulator in Drosophila. Cell Metab. 1, 323–330.

Guo, F., Yi, W., Zhou, M., Guo, A., 2011. Go signaling in mushroom bodies regulates sleep in Drosophila. Sleep 34, 273–281.

Guo, F., Yu, J., Jung, H.J., Abruzzi, K.C., Luo, W., Griffith, L.C., Rosbash, M., 2016. Circadian neuron feedback controls the Drosophila sleep—activity profile. Nature 536, 292–297.

Guo, F., Holla, M., Díaz, M.M., Rosbash, M., 2018. A circadian output circuit controls sleep-wake arousal in Drosophila. Neuron 100, 624–635.e624.

Haber, A.L., Biton, M., Rogel, N., Herbst, R.H., Shekhar, K., Smillie, C., Burgin, G., Delorey, T.M., Howitt, M.R., Katz, Y., Tirosh, I., Beyaz, S., Dionne, D., Zhang, M., Raychowdhury, R., Garrett, W.S., Rozenblatt-Rosen, O., Shi, H.N., Yilmaz, O., Xavier, R.J., Regev, A., 2017. A single-cell survey of the small intestinal epithelium. Nature 551, 333–339.

Han, C., Jan, L.Y., Y. N., Jan, 2011. Enhancer-driven membrane markers for analysis of non-autonomous mechanisms reveal neuron-glia interactions in Drosophila. Proc. Natl. Acad. Sci. U. S. A. 108, 9673–9678.

Harbison, S.T., McCoy, L.J., Mackay, T.F., 2013. Genome-wide association study of sleep in Drosophila melanogaster. BMC Genomics 14, 281.

Harbison, S.T., Serrano Negron, Y.L., Hansen, N.F., Lobell, A.S., 2017. Selection for long and short sleep duration in Drosophila melanogaster reveals the complex genetic network underlying natural variation in sleep. PLoS Genet. 13, e1007098.

Harmar, A.J., Marston, H.M., Shen, S., Spratt, C., West, K.M., Sheward, W.J., Morrison, C.F., Dorin, J.R., Piggins, H.D., Reubi, J.C., Kelly, J.S., Maywood, E.S., Hastings, M.H., 2002. The VPAC(2) receptor is essential for circadian function in the mouse suprachiasmatic nuclei. Cell 109, 497–508.

Hasler, G., Buysse, D.J., Klaghofer, R., Gamma, A., Ajdacic, V., Eich, D., Rössler, W., Angst, J., 2004. The association between short sleep duration and obesity in young adults: a 13-year prospective study. Sleep 27, 661–666.

Hayashi, Y., Nishimune, H., Hozumi, K., Saga, Y., Harada, A., Yuzaki, M., Iwatsubo, T., Kopan, R., Tomita, T., 2016. A novel non-canonical notch signaling regulates expression of synaptic vesicle proteins in excitatory neurons. Sci. Rep. 6, 23969.

Haynes, P.R., Christmann, B.L., Griffith, L.C., 2015. A single pair of neurons links sleep to memory consolidation in Drosophila melanogaster. Elife 4, e03868.

He, C.X., Yang, Y.Y., Zhang, M.M., Price, J.L., Zhao, Z.W., 2013. Regulation of sleep by neuropeptide Y-like system in Drosophila melanogaster. Plos One 8, e74237.

He, Q., Du, J., Wei, L., Zhao, Z., 2020. AKH-FOXO pathway regulates starvation-induced sleep loss through remodeling of the small ventral lateral neuron dorsal projections. PLoS Genet. 16, e1009181.

Hendricks, J.C., Finn, S.M., Panckeri, K.A., Chavkin, J., Williams, J.A., Sehgal, A., Pack, A.I., 2000. Rest in Drosophila is a sleep-like state. Neuron 25, 129–138.

Hendricks, J.C., Williams, J.A., Panckeri, K., Kirk, D., Tello, M., Yin, J.C., Sehgal, A., 2001. A non-circadian role for cAMP signaling and CREB activity in Drosophila rest homeostasis. Nat. Neurosci. 4, 1108–1115.

Holst, S.C., Landolt, H.-P., 2018. Sleep-wake neurochemistry. Sleep Med. Clin. 13, 137–146.

Huang, H., Zhu, C.T., Skuja, L.L., Hayden, D.J., Hart, A.C., 2017. Genome-wide screen for genes involved in Caenorhabditis elegans developmentally timed sleep. G3 (Bethesda) 7, 2907–2917.

Hunter-Ensor, M., Ousley, A., Sehgal, A., 1996. Regulation of the Drosophila protein timeless suggests a mechanism for resetting the circadian clock by light. Cell 84, 677–685.
Huntzinger, E., Izaurralde, E., 2011. Gene silencing by microRNAs: contributions of translational repression and mRNA decay. Nat. Rev. Genet. 12, 99–110.
Inutsuka, A., Yamanaka, A., 2013. The physiological role of orexin/hypocretin neurons in the regulation of sleep/wakefulness and neuroendocrine functions. Front. Endocrinol. (Lausanne) 4, 18.
Isaac, S.O., Berridge, C.W., 2003. Wake-promoting actions of dopamine D1 and D2 receptor stimulation. J. Pharmacol. Exp. Ther. 307, 386–394.
Isabel, G., Martin, J.-R., Chidami, S., Veenstra, J.A., Rosay, P., 2005. AKH-producing neuroendocrine cell ablation decreases trehalose and induces behavioral changes in Drosophila. Am. J. Physiol. Regul. Integr. Comp. Physiol. 288, R531.
Ishimoto, H., Kitamoto, T., 2010. The steroid molting hormone Ecdysone regulates sleep in adult Drosophila melanogaster. Genetics 185, 269–281.
Ito, S., Ueda, T., Ueno, A., Nakagawa, H., Taniguchi, H., Kayukawa, N., Miki, T., 2014. A genetic screen in Drosophila for regulators of human prostate cancer progression. Biochem. Biophys. Res. Commun. 451, 548–555.
Ito, K., Kawasaki, H., Suzuki, T., Takahara, T., Ishida, N., 2017. Effects of Kamikihito and Unkei-to on sleep behavior of wild type and Parkinson model in Drosophila. Front. Psych. 8, 132.
John, D., Berg, D.K., 2015. Long-lasting changes in neural networks to compensate for altered nicotinic input. Biochem. Pharmacol. 97, 418–424.
Joiner, W.J., Crocker, A., White, B.H., Sehgal, A., 2006. Sleep in drosophila is regulated by adult mushroom bodies. Nature 441, 757–760.
Juneau, B.A., Stonemetz, J.M., Toma, R.F., Possidente, D.R., Heins, R.C., Vecsey, C.G., 2019. Optogenetic activation of short neuropeptide F (sNPF) neurons induces sleep in Drosophila melanogaster. Physiol. Behav. 206, 143–156.
Kannangara, J.R., Henstridge, M.A., Parsons, L.M., Kondo, S., Mirth, C.K., Warr, C.G., 2020. A new role for neuropeptide F Signaling in controlling developmental timing and body size in Drosophila melanogaster. Genetics 216, 135–144.
Keene, A.C., Duboué, E.R., McDonald, D.M., Dus, M., Suh, G.S., Waddell, S., Blau, J., 2010. Clock and cycle limit starvation-induced sleep loss in Drosophila. Curr. Biol. 20, 1209–1215.
Kim, M., Lee, H., Hur, J.-H., Choe, J., Lim, C., 2016. CRTC potentiates light-independent timeless transcription to sustain circadian rhythms in Drosophila. Sci. Rep. 6, srep32113.
Kim, J.H., Ki, Y., Lee, H., Hur, M.S., Baik, B., Hur, J.H., Nam, D., Lim, C., 2020. The voltage-gated potassium channel shaker promotes sleep via thermosensitive GABA transmission. Commun. Biol. 3, 174.
Koh, K., Zheng, X., Sehgal, A., 2006. JETLAG resets the drosophila circadian clock by promoting light-induced degradation of TIMELESS. Science 312, 1809–1812.
Konopka, R.J., Benzer, S., 1971. Clock mutants of Drosophila melanogaster. Proc. Natl. Acad. Sci. U. S. A. 68, 2112–2116.
Krzemień, J., Dubois, L., Makki, R., Meister, M., Vincent, A., Crozatier, M., 2007. Control of blood cell homeostasis in Drosophila larvae by the posterior signalling centre. Nature 446, 325–328.
Kume, K., Kume, S., Park, S.K., Hirsh, J., Jackson, F.R., 2005. Dopamine is a regulator of arousal in the fruit fly. J. Neurosci. 25, 7377–7384.
Kunst, M., Hughes, M.E., Raccuglia, D., Felix, M., Li, M., Barnett, G., Duah, J., Nitabach, M.N., 2014. Calcitonin gene-related peptide neurons mediate sleep-specific circadian output in Drosophila. Curr. Biol. 24, 2652–2664.
Kurtishi, A., Rosen, B., Patil, K.S., Alves, G.W., Moller, S.G., 2019. Cellular proteostasis in neurodegeneration. Mol. Neurobiol. 56, 3676–3689.

Laposky, A.D., Shelton, J., Bass, J., Dugovic, C., Perrino, N., Turek, F.W., 2006. Altered sleep regulation in leptin-deficient mice. Am. J. Physiol. Regul. Integr. Comp. Physiol. 290, R894–R903.

Lazareva, A.A., Roman, G., Mattox, W., Hardin, P.E., Dauwalder, B., 2007. A role for the adult fat body in Drosophila male courtship behavior. PLoS Genet. 3, e16.

Lee, T.M., Luo, L.Q., 2001. Mosaic analysis with a repressible cell marker (MARCM) for Drosophila neural development. Trends Neurosci. 24, 251–254.

Levine, B.D., Cagan, R.L., 2016. Drosophila lung cancer models identify trametinib plus statin as candidate therapeutic. Cell Rep. 14, 1477–1487.

Li, Y., Zhou, Z., Zhang, X., Tong, H., Li, P., Zhang, Z.C., Jia, Z., Xie, W., Han, J., 2013. Drosophila neuroligin 4 regulates sleep through modulating GABA transmission. J. Neurosci. 33, 15545–15554.

Li, Q., Li, Y., Wang, X., Qi, J., Jin, X., Tong, H., Zhou, Z., Zhang, Z.C., Han, J., 2017. Fbxl4 serves as a clock output molecule that regulates sleep through promotion of rhythmic degradation of the GABA(A) receptor. Curr. Biol. 27, 3616.

Li, L., Zhang, L., Yang, C.C., 2016. Multi-target strategy and experimental studies of traditional Chinese medicine for Alzheimer's disease therapy. Curr. Top. Med. Chem. 16, 537–548.

Lin, F.J., Song, W., Meyer-Bernstein, E., Naidoo, N., Sehgal, A., 2001. Photic signaling by cryptochrome in the Drosophila circadian system. Mol. Cell. Biol. 21, 7287–7294.

Liu, Q., Liu, S., Kodama, L., Driscoll, M.R., Wu, M.N., 2012. Two dopaminergic neurons signal to the dorsal fan-shaped body to promote wakefulness in Drosophila. Curr. Biol. 22, 2114–2123.

Liu, S., Lamaze, A., Liu, Q., Tabuchi, M., Yang, Y., Fowler, M., Bharadwaj, R., Zhang, J., Bedont, J., Blackshaw, S., Lloyd, T.E., Montell, C., Sehgal, A., Koh, K., Wu, M.N., 2014. WIDE AWAKE mediates the circadian timing of sleep onset. Neuron 82, 151–166.

Liu, S., Liu, Q., Tabuchi, M., Wu, M.N., 2016. Sleep drive is encoded by neural plastic changes in a dedicated circuit. Cell 165, 1347–1360.

Lu, J., Jhou, T.C., Saper, C.B., 2006. Identification of wake-active dopaminergic neurons in the ventral periaqueductal gray matter. J. Neurosci. 26, 193–202.

Ly, S., Pack, A.I., Naidoo, N., 2018. The neurobiological basis of sleep: insights from Drosophila. Neurosci. Biobehav. Rev. 87, 67–86.

Ly, S., Lee, D.A., Strus, E., Prober, D.A., Naidoo, N., 2020. Evolutionarily conserved regulation of sleep by the protein translational regulator PERK. Curr. Biol. 30, 1639–1648.

Makos, M.A., Kim, Y.-C., Han, K.-A., Heien, M.L., Ewing, A.G., 2009. In vivo electrochemical measurements of exogenously applied dopamine in Drosophila melanogaster. Anal. Chem. 81, 1848–1854.

Mao, Z., Davis, R.L., 2009. Eight different types of dopaminergic neurons innervate the Drosophila mushroom body neuropil: anatomical and physiological heterogeneity. Front. Neural Circuits 3, 1–17.

Maurer, C., Winter, T., Chen, S., Hung, H.C., Weber, F., 2016. The CREB-binding protein affects the circadian regulation of behaviour. FEBS Lett. 590, 3213–3220.

Mayford, M., Kandel, E.R., 1999. Genetic approaches to memory storage. Trends Genet. 15, 463–470.

McCarthy, E.V., Wu, Y., deCarvalho, T., Brandt, C., Cao, G., Nitabach, M.N., 2011. Synchronized bilateral synaptic inputs to Drosophila melanogaster neuropeptidergic rest/arousal neurons. J. Neurosci. 31, 8181–8193.

Menon, K.P., Sanyal, S., Habara, Y., Sanchez, R., Wharton, R.P., Ramaswami, M., Zinn, K., 2004. The translational repressor Pumilio regulates presynaptic morphology and controls postsynaptic accumulation of translation factor eIF-4E. Neuron 44, 663–676.

Mercer, R.E., Chee, M.J.S., Colmers, W.F., 2011. The role of NPY in hypothalamic mediated food intake. Front. Neuroendocrinol. 32, 398–415.

Metaxakis, A., Tain, L.S., Nke, S.G., Hendrich, O., Hinze, Y., 2014. Lowered insulin signalling ameliorates age-related sleep fragmentation in *Drosophila*. PLoS Biol. 12, e1001824.

Mezler, M., Muller, T., Raming, K., 2001. Cloning and functional expression of GABA(B) receptors from Drosophila. Eur. J. Neurosci. 13, 477–486.

Missale, C., Nash, S.R., Robinson, S.W., Jaber, M., Caron, M.G., 1998. Dopamine receptors: from structure to function. Physiol. Rev. 78, 189–225.

Morton, G.J., Schwartz, M.W., 2011. Leptin and the central nervous system control of glucose metabolism. Physiol. Rev. 91, 389–411.

Murakami, K., Yurgel, M.E., Stahl, B.A., Masek, P., Mehta, A., Heidker, R., Bollinger, W., Gingras, R.M., Kim, Y.J., Ja, W.W., Suter, B., DiAngelo, J.R., Keene, A.C., 2016. Translin is required for metabolic regulation of sleep. Curr. Biol. 26, 972–980.

Myers, M.P., Wager-Smith, K., Rothenfluh-Hilfiker, A., Young, M.W., 1996. Light-induced degradation of TIMELESS and entrainment of the Drosophila circadian clock. Science 271, 1736–1740.

Naidoo, N., 2009. Cellular stress/the unfolded protein response: relevance to sleep and sleep disorders. Sleep Med. Rev. 13, 195–204.

Naidoo, N., Song, W., Hunter-Ensor, M., Sehgal, A., 1999. A role for the proteasome in the light response of the timeless clock protein. Science 285, 1737–1741.

Naidoo, N., Casiano, V., Cater, J., Zimmerman, J., Pack, A.I., 2007. A role for the molecular chaperone protein BiP/GRP78 in Drosophila sleep homeostasis. Sleep 30, 557–565.

Ni, J.D., Gurav, A.S., Liu, W., Ogunmowo, T.H., Hackbart, H., Elsheikh, A., Verdegaal, A.A., Montell, C., 2019. Differential regulation of the Drosophila sleep homeostat by circadian and arousal inputs. Elife 8, e40487.

Oh, Y., Jang, D., Sonn, J.Y., Choe, J., 2013. Histamine-HisCl1 receptor axis regulates wake-promoting signals in Drosophila melanogaster. Plos One 8, e68269.

Oh, Y., Yoon, S.E., Zhang, Q., Chae, H.S., Daubnerova, I., Shafer, O.T., Choe, J., Kim, Y.J., 2014. A homeostatic sleep-stabilizing pathway in Drosophila composed of the sex peptide receptor and its ligand, the myoinhibitory peptide. PLoS Biol. 12, e1001974.

Oikonomou, G., Altermatt, M., Zhang, R.W., Coughlin, G.M., Montz, C., Gradinaru, V., Prober, D.A., 2019. The serotonergic raphe promote sleep in zebrafish and mice. Neuron 103, 686–701.

Okada, R., Awasaki, T., Ito, K., 2009. Gamma-aminobuyric acid (GABA)-mediated neural connections in the drosophila antennal lobe. J. Comp. Neurol. 514, 74–91.

Osman, D., Gobert, V., Ponthan, F., Heidenreich, O., Haenlin, M., Waltzer, L., 2009. A Drosophila model identifies calpains as modulators of the human leukemogenic fusion protein AML1-ETO. Proc. Natl. Acad. Sci. U. S. A. 106, 12043–12048.

Paisán-Ruíz, C., Jain, S., Evans, E.W., Gilks, W.P., Simón, J., van der Brug, M., López de Munain, A., Aparicio, S., Gil, A.M., Khan, N., Johnson, J., Martinez, J.R., Nicholl, D., Martí Carrera, I., Pena, A.S., de Silva, R., Lees, A., Martí-Massó, J.F., Pérez-Tur, J., Wood, N.W., Singleton, A.B., 2004. Cloning of the gene containing mutations that cause PARK8-linked Parkinson's disease. Neuron 44, 595–600.

Pallier, P.N., Morton, A.J., 2009. Management of sleep/wake cycles improves cognitive function in a transgenic mouse model of Huntington's disease. Brain Res. 1279, 90–98.

Papouin, T., Dunphy, J.M., Tolman, M., Dineley, K.T., Haydon, P.G., 2017. Septal cholinergic neuromodulation tunes the astrocyte-dependent gating of hippocampal NMDA receptors to wakefulness. Neuron 94, 840–854.e847.

Parisky, K.M., Agosto, J., Pulver, S.R., Shang, Y.H., Kuklin, E., Hodge, J.J.L., Kang, K., Liu, X., Garrity, P.A., Rosbash, M., Griffith, L.C., 2008. PDF cells are a GABA-responsive wake-promoting component of the Drosophila sleep circuit. Neuron 60, 672–682.

Parisky, K.M., Agosto Rivera, J.L., Donelson, N.C., Kotecha, S., Griffith, L.C., 2016. Reorganization of sleep by temperature in Drosophila requires light, the homeostat, and the circadian clock. Curr. Biol. 26, 882–892.

Peschel, N., Helfrich-Foerster, C., 2011. Setting the clock—by nature: circadian rhythm in the fruitfly Drosophila melanogaster. FEBS Lett. 585, 1435–1442.

Pimentel, D., Donlea, J.M., Albot, C.B.T., Ong, S.M.S., Hurston, A., Miesenbock, G., 2016. Operation of a homeostatic sleep switch. Nature 536, 333–337.

Pirooznia, S.K., Chiu, K., Chan, M.T., Zimmerman, J.E., Elefant, F., 2012. Epigenetic regulation of axonal growth of Drosophila pacemaker cells by histone acetyltransferase tip60 controls sleep. Genetics 192, 1327–1345.

Pitman, J.L., McGill, J.J., Keegan, K.P., Allada, R., 2006. A dynamic role for the mushroom bodies in promoting sleep in Drosophila. Nature 441, 753–756.

Potdar, S., Sheeba, V., 2018. Wakefulness is promoted during day time by PDFR signalling to dopaminergic neurons in Drosophila melanogaster. eNeuro 5, e0129-0118.

Qian, Y., Cao, Y., Deng, B., Yang, G., Li, J., Xu, R., Zhang, D., Huang, J., Rao, Y., 2017. Sleep homeostasis regulated by 5HT2b receptor in a small subset of neurons in the dorsal fan-shaped body of Drosophila. Elife 6, e26519.

Qu, W.-M., Xu, X.-H., Yan, M.-M., Wang, Y.-Q., Urade, Y., Huang, Z.-L., 2010. Essential role of dopamine D-2 receptor in the maintenance of wakefulness, but not in homeostatic regulation of sleep, in mice. J. Neurosci. 30, 4382–4389.

Rajan, A., Perrimon, N., 2012. Drosophila cytokine unpaired 2 regulates physiological homeostasis by remotely controlling insulin secretion. Cell 151, 123–137.

Rogulja, D., Young, M.W., 2012. Control of sleep by cyclin A and its regulator. Science 335, 1617–1621.

Ron, D., Walter, P., 2007. Signal integration in the endoplasmic reticulum unfolded protein response. Nat. Rev. Mol. Cell Biol. 8, 519–529.

Rye, D.B., Jankovic, J., 2002. Emerging views of dopamine in modulating sleep/wake state from an unlikely source. PD. Neurology 58, 341–346.

Sajwan, S., Sidorov, R., Staskova, T., Zaloudikova, A., Takasu, Y., Kodrik, D., Zurovec, M., 2015. Targeted mutagenesis and functional analysis of adipokinetic hormone-encoding gene in *Drosophila*. Insect Biochem. Mol. Biol. 61, 79–86.

Salvaterra, P.M., Kitamoto, T., 2001. Drosophila cholinergic neurons and processes visualized with Gal4/UAS-GFP. Brain Res. Gene Expr. Patterns 1, 73–82.

Saper, C.B., Fuller, P.M., Pedersen, N.P., Lu, J., Scammell, T.E., 2010. Sleep state switching. Neuron 68, 1023–1042.

Sassu, E.D., McDermott, J.E., Keys, B.J., Esmaeili, M., Keene, A.C., Birnbaum, M.J., DiAngelo, J.R., 2012. Mio/dChREBP coordinately increases fat mass by regulating lipid synthesis and feeding behavior in Drosophila. Biochem. Biophys. Res. Commun. 426, 43–48.

Saudou, F., Boschert, U., Amlaiky, N., Plassat, J.L., Hen, R., 1992. A family of Drosophila serotonin receptors with distinct intracellular signaling properties and expression patterns. EMBO J. 11, 7–17.

Serrano Negron, Y.L., Hansen, N.F., Harbison, S.T., 2018. The sleep inbred panel, a collection of inbred *Drosophila melanogaster* with extreme long and short sleep duration. G3 (Bethesda) 8, 2865–2873.

Seugnet, L., Galvin, J.E., Suzuki, Y., Gottschalk, L., Shaw, P.J., 2009. Persistent short-term memory defects following sleep deprivation in a drosophila model of Parkinson disease. Sleep 32, 984–992.

Seugnet, L., Suzuki, Y., Merlin, G., Gottschalk, L., Duntley, S.P., Shaw, P.J., 2011. Notch signaling modulates sleep homeostasis and learning after sleep deprivation in Drosophila. Curr. Biol. 21, 835–840.

Seugnet, L., Dissel, S., Thimgan, M., Cao, L., Shaw, P.J., 2017. Identification of genes that maintain behavioral and structural plasticity during sleep loss. Front. Neural Circuits 11, 79.

Shang, Y., Griffith, L.C., Rosbash, M., 2008. Light-arousal and circadian photoreception circuits intersect at the large PDF cells of the Drosophila brain. Proc. Natl. Acad. Sci. U. S. A. 105, 19587–19594.

Shang, Y.H., Donelson, N.C., Vecsey, C.G., Guo, F., Rosbash, M., Griffith, L.C., 2013. Short neuropeptide F is a sleep-promoting inhibitory modulator. Neuron 80, 171–183.

Shaw, P.J., Cirelli, C., Greenspan, R.J., Tononi, G., 2000. Correlates of sleep and waking in Drosophila melanogaster. Science 287, 1834–1837.

Shaw, P.J., Tononi, G., Greenspan, R.J., Robinson, D.F., 2002. Stress response genes protect against lethal effects of sleep deprivation in Drosophila. Nature 417, 287–291.

Sheeba, V., Fogle, K.J., Kaneko, M., Rashid, S., Chou, Y.T., Sharma, V.K., Holmes, T.C., 2008. Large ventral lateral neurons modulate arousal and sleep in Drosophila. Curr. Biol. 18, 1537–1545.

Sinakevitch, I., Strausfeld, N.J., 2006. Comparison of octopamine-like immunoreactivity in the brains of the fruit fly and blow fly. J. Comp. Neurol. 494, 460–475.

Sinakevitch-Pean, I., Geffard, M., Plotnikova, S.I., 2001. Localization of glutamate in the nervous system of the fly Drosophila melanogaster: An immunocytochemical study. J. Evol. Biochem. Physiol. 37, 83–88.

Sinenko, S.A., Hung, T., Moroz, T., Tran, Q.M., Sidhu, S., Cheney, M.D., Speck, N.A., Banerjee, U., 2010. Genetic manipulation of AML1-ETO-induced expansion of hematopoietic precursors in a Drosophila model. Blood 116, 4612–4620.

Singh, K., Chao, M.Y., Somers, G.A., Komatsu, H., Corkins, M.E., Larkins-Ford, J., Tucey, T., Dionne, H.M., Walsh, M.B., Beaumont, E.K., Hart, D.P., Lockery, S.R., Hart, A.C., 2011. C. elegans notch signaling regulates adult chemosensory response and larval molting quiescence. Curr. Biol. 21, 825–834.

Singh, K., Ju, J.Y., Walsh, M.B., DiIorio, M.A., Hart, A.C., 2014. Deep conservation of genes required for both Drosphila melanogaster and Caenorhabditis elegans sleep includes a role for dopaminergic signaling. Sleep 37, 1439–1451.

Sitaraman, D., Aso, Y., Rubin, G.M., Nitabach, M.N., 2015. Control of sleep by dopaminergic inputs to the Drosophila mushroom body. Front. Neural Circuits 9, 73.

Sokol, N.S., Xu, P., Jan, Y.N., Ambros, V., 2008. Drosophila let-7 microRNA is required for remodeling of the neuromusculature during metamorphosis. Genes Dev. 22, 1591–1596.

Song, Q., Feng, G., Huang, Z., Chen, X., Chen, Z., Ping, Y., 2017. Aberrant axonal arborization of PDF neurons induced by Abeta42-mediated JNK activation underlies sleep disturbance in an Alzheimer's model. Mol. Neurobiol. 54, 6317–6328.

Sonn, J.Y., Lee, J., Sung, M.K., Ri, H., Choi, J.K., Lim, C., Choe, J., 2018. Serine metabolism in the brain regulates starvation-induced sleep suppression in *Drosophila melanogaster*. Proc. Natl. Acad. Sci. U. S. A. 115, 7129–7134.

Stanewsky, R., Kaneko, M., Emery, P., Beretta, B., Wager-Smith, K., Kay, S.A., Rosbash, M., Hall, J.C., 1998. The cryb mutation identifies cryptochrome as a circadian photoreceptor in Drosophila. Cell 95, 681–692.

Stavropoulos, N., Young, M.W., 2011. Insomniac and Cullin-3 regulate sleep and wakefulness in Drosophila. Neuron 72, 964–976.

Sun, X., Ran, D., Zhao, X., Huang, Y., Long, S., Liang, F., Guo, W., Nucifora Jr., F.C., Gu, H., Lu, X., Chen, L., Zeng, J., Ross, C.A., Pei, Z., 2016. Melatonin attenuates hLRRK2-induced sleep disturbances and synaptic dysfunction in a Drosophila model of Parkinson's disease. Mol. Med. Rep. 13, 3936–3944.

Szerb, J.C., 1967. Cortical acetylcholin release and electroencephalographic arousal. J. Physiol-London 192, 329.

Takahama, K., Tomita, J., Ueno, T., Yamazaki, M., Kume, S., Kume, K., 2012. Pan-neuronal knockdown of the c-Jun N-terminal kinase (JNK) results in a reduction in sleep and longevity in Drosophila. Biochem. Biophys. Res. Commun. 417, 807–811.

Thimgan, M.S., Suzuki, Y., Seugnet, L., Gottschalk, L., Shaw, P.J., 2010. The perilipin homologue, lipid storage droplet 2, regulates sleep homeostasis and prevents learning impairments following sleep loss. PLoS Biol. 8, e1000466.

Tomita, J., Ueno, T., Mitsuyoshi, M., Kume, S., Kume, K., 2015. The NMDA receptor promotes sleep in the fruit Fly, Drosophila melanogaster. Plos One 10, e0128101.

Ueno, T., Tomita, J., Tanimoto, H., Endo, K., Ito, K., Kume, S., Kume, K., 2012. Identification of a dopamine pathway that regulates sleep and arousal in drosophila. Nat. Neurosci. 15, 1516–1523.

Ursin, R., 2002. Serotonin and sleep. Sleep Med. Rev. 6, 57–69.

Vaccaro, A., Kaplan Dor, Y., Nambara, K., Pollina, E.A., Lin, C., Greenberg, M.E., Rogulja, D., 2020. Sleep loss can cause death through accumulation of reactive oxygen species in the gut. Cell 181, 1307–1328.e1315.

Valadas, J.S., Esposito, G., Vandekerkhove, D., Miskiewicz, K., Deaulmerie, L., Raitano, S., Seibler, P., Klein, C., Verstreken, P., 2018. ER lipid defects in neuropeptidergic neurons impair sleep patterns in Parkinson's disease. Neuron 98, 1155–1169.e1156.

Van Dort, C.J., Zachs, D.P., Kenny, J.D., Zheng, S., Goldblum, R.R., Gelwan, N.A., Ramos, D.M., Nolan, M.A., Wang, K., Weng, F.-J., Lin, Y., Wilson, M.A., Brown, E.N., 2015. Optogenetic activation of cholinergic neurons in the PPT or LDT induces REM sleep. Proc. Natl. Acad. Sci. U. S. A. 112, 584–589.

Vazquez, J., Baghdoyan, H.A., 2001. Basal forebrain acetylcholine release during REM sleep is significantly greater than during waking. Am. J. Physiol. Regul. Integr. Comp. Physiol. 280, R598–R601.

Vessey, J.P., Vaccani, A., Xie, Y., Dahm, R., Karra, D., Kiebler, M.A., Macchi, P., 2006. Dendritic localization of the translational repressor Pumilio 2 and its contribution to dendritic stress granules. J. Neurosci. 26, 6496–6508.

Vodala, S., Pescatore, S., Rodriguez, J., Buescher, M., Chen, Y.-W., Weng, R., Cohen, S.M., Rosbash, M., 2012. The oscillating miRNA 959-964 cluster impacts Drosophila feeding time and other circadian outputs. Cell Metab. 16, 601–612.

Volkow, N.D., Fowler, J.S., Logan, J., Alexoff, D., Zhu, W., Telang, F., Wang, G.-J., Jayne, M., Hooker, J.M., Wong, C., Hubbard, B., Carter, P., Warner, D., King, P., Shea, C., Xu, Y., Muench, L., Apelskog-Torres, K., 2009. Effects of Modafinil on dopamine and dopamine transporters in the male human brain clinical implications. JAMA 301, 1148–1154.

Wang, C., Guo, X., Xi, R., 2014. EGFR and notch signaling respectively regulate proliferative activity and multiple cell lineage differentiation of Drosophila gastric stem cells. Cell Res. 24, 610–627.

Weston, A.J., Baines, R.A., 2007. Translational regulation of neuronal electrical properties. Invert. Neurosci. 7, 75–86.

Wilson, C., Leiblich, A., Goberdhan, D.C., Hamdy, F., 2017. The Drosophila accessory gland as a model for prostate cancer and other pathologies. Curr. Top. Dev. Biol. 121, 339–375.

Wisor, J.P., Nishino, S., Sora, I., Uhl, G.H., Mignot, E., Edgar, D.M., 2001. Dopaminergic role in stimulant-induced wakefulness. J. Neurosci. 21, 1787–1794.

Witz, P., Amlaiky, N., Plassat, J.L., Maroteaux, L., Borrelli, E., Hen, R., 1990. Cloning and characterization of a Drosophila serotonin receptor that activates adenylate-cyclase. Proc. Natl. Acad. Sci. U. S. A. 87, 8940–8944.

Wu, B.B., Ma, L.L., Zhang, E.Y., Du, J., Liu, S.N., Price, J., Li, S., Zhao, Z.W., 2018. Sexual dimorphism of sleep regulated by juvenile hormone signaling in Drosophila. PLoS Genet. 14, e1007318.

Xia, X.X., Fu, J., Du, B., Wu, X., Zhao, J.Z., Zhao, Z., 2020. Regulation of circadian rhythm and sleep by miR-375-timeless interaction in Drosophila. FASEB J. 34, 16536–16551.

Xu, K., DiAngelo, J.R., Hughes, M.E., Hogenesch, J.B., Sehgal, A., 2011. The circadian clock interacts with metabolic physiology to influence reproductive fitness. Cell Metab. 13, 639–654.

Xiong, M., Li, J., Wang, D., Delphin, E., Ye, J.H., 2012. Intra-ventrolateral preoptic nucleus injection of gamma-aminobutyric acid induces sedation in rats. Int. J. Physiol. Pathophysiol. Pharmacol. 4, 94–98.

Xu, M., Chung, S., Zhang, S., Zhong, P., Ma, C., Chang, W.-C., Weissbourd, B., Sakai, N., Luo, L., Nishino, S., Dan, Y., 2015. Basal forebrain circuit for sleep-wake control. Nat. Neurosci. 18, 1641–1647.

Yap, M.H.W., Grabowska, M.J., Rohrscheib, C., Jeans, R., Troup, M., Paulk, A.C., van Alphen, B., Shaw, P.J., van Swinderen, B., 2017. Oscillatory brain activity in spontaneous and induced sleep stages in flies. Nat. Commun. 8, 1815.

Yasuyama, K., Salvaterra, P.M., 1999. Localization of choline acetyltransferase-expressing neurons in Drosophila nervous system. Microsc. Res. Tech. 45, 65–79.

Yi, W., Zhang, Y., Tian, Y., Guo, J., Li, Y., Guo, A., 2013. A subset of cholinergic mushroom body neurons requires go signaling to regulate sleep in Drosophila. Sleep 36, 1809–1821.

Yokoi, T., Nabe, T., Ishizuka, C., Hayashi, K., Ito-Harashima, S., Yagi, T., Nakagawa, Y., Miyagawa, H., 2020. Transcription-inducing activity of natural and synthetic juvenile hormone agonists through the Drosophila Methoprene-tolerant protein. Pest Manag. Sci. 76, 2316–2323.

Yuan, L.-L., Chen, X., 2006. Diversity of potassium channels in neuronal dendrites. Prog. Neurobiol. 78, 374–389.

Yuan, Q., Joiner, W.J., Sehgal, A., 2006. A sleep-promoting role for the Drosophila serotonin receptor 1A. Curr. Biol. 16, 1051–1062.

Yuan, X., Wu, H., Xu, H., Xiong, H., Chu, Q., Yu, S., Wu, G.S., Wu, K., 2015. Notch signaling: an emerging therapeutic target for cancer treatment. Cancer Lett. 369, 20–27.

Yurgel, M.E., Kakad, P., Zandawala, M., Nassel, D.R., Godenschwege, T.A., Keene, A.C., 2019. A single pair of leucokinin neurons are modulated by feeding state and regulate sleep-metabolism interactions. PLoS Biol. 17, e2006409.

Zamboni, G., Perez, E., Amici, R., Jones, C.A., Parmeggiani, P.L., 1999. Control of REM sleep: an aspect of the regulation of physiological homeostasis. Arch. Ital. Biol. 137, 249–262.

Zhang, X., Yan, H.M., Luo, Y.J., Huang, Z.L., Rao, Y., 2018. Thermoregulation-independent regulation of sleep by serotonin revealed in mice defective in serotonin synthesis. Mol. Pharmacol. 93, 657–664.

Zhang, R., Du, J., Zhao, X., Wei, L., Zhao, Z., 2020. Regulation of circadian behavioural output via clock-responsive miR-276b. Insect Mol. Biol. 30, 81–85.

Zhang, R., Zhao, X., Du, J., Wei, L., Zhao, Z., 2021. Regulatory mechanism of daily sleep by miR-276a. FASEB J. 35, e21222.

Zhao, X., Li, Y., Zhao, Z., Du, J., 2021a. Extra sex combs buffers sleep-related stresses through regulating heat shock proteins. FASEB J. 35, e21190.

Zhao, Z., Zhao, X., He, T., Wu, X., Lv, P., Zhu, A.J., Du, J., 2021b. Epigenetic regulator Stuxnet modulates octopamine effect on sleep through a Stuxnet-Polycomb-Octb2R cascade. EMBO Rep. 22, e47910.

Zimmerman, J.E., Chan, M.T., Lenz, O.T., Keenan, B.T., Maislin, G., Pack, A.I., 2017. Glutamate is a wake-active neurotransmitter in Drosophila melanogaster. Sleep 40, zsw046.

FSC
www.fsc.org
MIX
Papier | Fördert
gute Waldnutzung
FSC® C083411

Kolibri 360 GmbH
Ferdinand-Jühlke-Straße 7
99095 Erfurt, Deutschland
produktsicherheit@kolibri360.de